CONNECTING CARE

CONNECTING CARE

UNLOCKING THE POTENTIAL OF ELECTRONIC HEALTH RECORDS TO UNITE PATIENTS & PROVIDERS

KEN HOFFMAN | GILBERT PANT

Advantage | Books

Published by Advantage Books, Charleston, South Carolina.
An imprint of Advantage Media.

ADVANTAGE is a registered trademark, and the Advantage colophon is a trademark of Advantage Media Group, Inc.

Printed in the United States of America.

10 9 8 7 6 5 4 3 2 1

ISBN: 978-1-64225-598-0 (Paperback)
ISBN: 978-1-64225-597-3 (eBook)

Library of Congress Control Number: 2024911068

Cover design by Megan Elger.
Layout design by Ruthie Wood.

This publication is designed to provide accurate and authoritative information in regard to the subject matter covered. It is sold with the understanding that the publisher is not engaged in rendering legal, accounting, or other professional services. If legal advice or other expert assistance is required, the services of a competent professional person should be sought.

Advantage Books is an imprint of Advantage Media Group. Advantage Media helps busy entrepreneurs, CEOs, and leaders write and publish a book to grow their business and become the authority in their field. Advantage authors comprise an exclusive community of industry professionals, idea-makers, and thought leaders. For more information go to **advantagemedia.com**.

CONTENTS

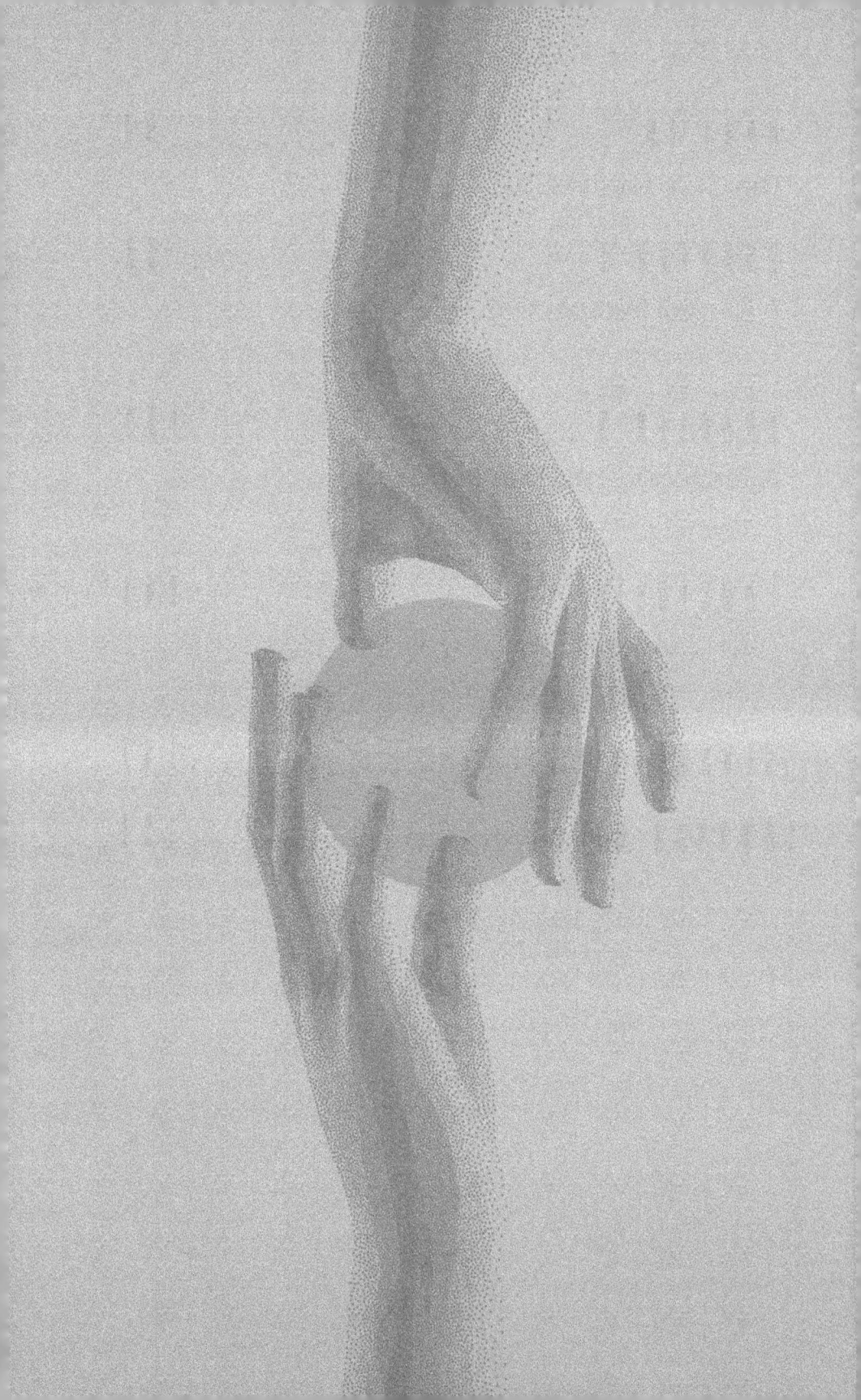

PROLOGUE

Healthcare has been an absolute requirement for national
defense and for a productive healthy workforce.

Over the past fifty years—reflecting on my career through medical
education, residency, and practice—there has been a massive transformation in both healthcare and the use of technology, including the
transitions from paper medical records to electronic health records
(EHRs). More so today than before, I question the primary purpose
for documentation, which was once so essential for good medical care.

A Clinical Rationale for Documentation

In the March 14, 1968, issue of the *New England Journal of Medicine*,
Larry Weed, MD—the father of the "SOAP" (Subjective, Objective,
Assessment, and Plan) note—wrote a special article titled "Medical
Records That Guide and Teach." The article opens with:

"It has been said that preoccupation with the medical record
and computer leads to neglect of the 'humanitarian' side and the 'art'
of medicine. The most humanitarian thing a physician can do is to

precisely know what he is doing and make the patient as comfortable as he can in the face of problems that he cannot yet solve."

For the past thirty years, I have been acutely aware of the tremendous amount of person-years and costs that have gone into strategic plans and business case analyses to generate the functional requirements organizations use to contract for EHRs, just to replace paper medical records. There is always the expectation that with documentation now in an EHR, it will improve the quality of care. With increasing computing capability to collect and store vast amounts of data, somehow the potential for technology to enhance the quality of care is being lost. In the strangest development, EHRs are now more likely to be perceived as serving the needs of healthcare administrators for patient scheduling and billing. EHRs have not been designed to help improve the quality of a patient encounter or the decision-making process critical for an accurate diagnosis, and they have not improved a shared decision-making process necessary for an individualized and effective treatment plan. "Concurrent documentation," once routine for paper records, has become increasingly difficult, leading to delayed documentation and requiring clinicians to donate hours after work to meet the needs of the EHR.

This current definition from the Office of the National Coordinator for Health Information Technology (ONC) remains remarkably consistent with what was envisioned in 1968.

"An electronic health record (EHR) is a digital version of a patient's paper chart. EHRs are real-time, patient-centered records that make information available instantly and securely to authorized users. While an EHR does contain the medical and treatment histories of patients, an EHR system is built to go beyond standard clinical data collected in a provider's office and can be inclusive of a broader view of a patient's care. EHRs are a vital part of health IT and can:

- Contain a patient's medical history, diagnoses, medications, treatment plans, immunization dates, allergies, radiology images, and laboratory and test results.

- Allow access to evidence-based tools that providers can use to make decisions about a patient's care.

- Automate and streamline provider workflow.

One of the key features of an EHR is that health information can be created and managed by authorized providers in a digital format capable of being shared with other providers across more than one healthcare organization. EHRs are built to share information with other healthcare providers and organizations—such as laboratories, specialists, medical imaging facilities, pharmacies, emergency facilities, and school and workplace clinics—so they contain information from *all clinicians involved in a patient's care.*"[1]

Once heavily engaged—and now reading about the benefits of EHRs through government, industry, and healthcare websites   I'm faced with the reality that for each patient I see, the EHR has become an overarching presence, competing for my time and focus that I'd rather dedicate to the patient in front of me. Completing a note while the patient is still with me has become increasingly difficult as I try to find workarounds. While vast amounts of data are in the EHR, so little is of practical use within the patient encounter, yet critical information can be easily missed in the mass amount of unindexed data. While I went into medicine to help others, I now find myself dreading the EHR that seems purposefully designed for clinician burnout. The EHR I'm required to use, rarely, if ever, imports critical data from

---

1   "What Is An Electronic Health Record (EHR)?" HealthIT.gov, accessed April 16, 2024, https://www.healthit.gov/faq/what-electronic-health-record-ehr.

other places the patient has been or from other providers involved in the patient's care. It doesn't even alert me that such data exists.

Now in the twilight of my career, the vision I had for a patient-centered EHR, once a tangible reality in a project I directed, seems no longer possible. I never thought my experience may have been unique: an optimization business process reengineering study that combined a concurrent "top-down" analysis of a national and Department of Defense (DOD) vision for population health and "bottom-up" analysis of clinicians at work taking care of patients. We defined the clinical work necessary to achieve the vision and eliminate labor that added no value but had cost. Designing an EHR that would help achieve that vision through optimal patient outcomes was integral to that project.

Throughout this book, I will use a very specific example: the Army's alcohol and drug prevention and control program which operated as its own medical specialty, by public law and DOD directives in response to the opioid epidemic occurring with our troops in Vietnam. This was identified as a national concern in 1969. Combined with a general level of problematic alcohol use throughout the military, public law was passed in 1971, continuing today that secretaries of all agencies should be covered under TRICARE. This plan was meant to "prescribe regulations, implement procedures using each practical and available method, and provide necessary facilities to identify, treat, and rehabilitate members of the armed forces who are dependent on drugs or alcohol" (Title 10 US Code Chapter 55 section 1090).

Mission, Vision, and Purpose of Healthcare

In the 1880s in Germany, Otto von Bismarck, with a goal to increase worker productivity and preserve capitalism from the rise of socialism, advocated the concept of "practical Christianity" to describe state programs that provide the security needed by workers who might become ill or disabled or have an accident that would result in unemployment and poverty. Seeing a need and obligation to help those who had worked their entire lives, Bismarck led passage through the Reichstag of three bills: the Health Insurance Bill of 1883, the Accident Insurance Bill of 1884, and the Old Age and Disability Bill of 1889. National healthcare, disability, and retirement systems have evolved from the concepts of these bills. This was the genesis of national healthcare systems within most developed nations outside the United States today.

In a European healthcare environment, Gilbert Pant saw the need for an EHR designed to document each patient encounter in whatever manner the clinician found most helpful, in any language spoken by the clinician, while in the background notes were coded into international standard vocabularies such as "SNOMED-CT." This allowed connections between clinics across the world and generated reports that would be useful for outcome management and population health. The Korea Case study highlights the synergy possible when IT supports population health improvement.

Medical care has always been integral to the US military since the days of the Militia Act of 1792. A medical officer was part of the command staff. At that time, US male citizens were required to serve in the militia, and all needed to be fit and ready to defend the country. If ill or injured, they would receive the treatment necessary to return to duty when possible or to be sent home in as good of a condition as possible. Medical care was important for morale—that if ill or injured, the best medical care possible would be provided. In public law since 1958, the purpose of having a military healthcare system is "to create and maintain high morale in the uniformed services by providing an improved and uniform program of medical and dental care for members and certain former members of those services, and for their dependents" (Title 10 US Code Chapter 55 section 1071).

In 1998, lessons learned from the first Gulf War and deployments in Bosnia, Haiti, Rwanda, and Somalia led to the publication of the Presidential Review Directive 5, titled "A National Obligation: Planning for Health Preparedness for and Readjustment of the Military, Veterans, and Their Families after Future Deployments." Goals and objectives were defined to better address health concerns, healthcare, health risk communication, recordkeeping, and coordinating research designed to fill in gaps in knowledge that could be applied to health improvement.

What I realized today is that the approach we describe in chapter 1 may be a unique application coming out of a program that operated independently of the usual hospital programs—more community-based and covering the spectrum from primary to tertiary prevention. In 1991, I received a grant through the Army Medical Command to create a Center for Training and Education in Addiction Medicine (CTEAM) at the Uniformed Services University. I was given the opportunity to work closely with the Army's drug and alcohol prevention and treatment programs, which existed at every Army installation worldwide, and the Tri-service Alcohol Rehabilitation Department at National Navy Medical Center.

Between 1980 and 2000, the US military developed a healthcare model with seven steps for population health improvement.

There was a critical link between the healthy community and the medical/military health system (MHS). Determinants of health were an interactive relationship between the individual, other people, and the surrounding physical environment. The individual could be defined in data in terms of both biological and behavioral factors. Policies and interventions related to health within the community included access to medical care.

The seven steps for population health improvement rely on having data for each step. (1) Identify the population. This generally would come from personnel systems identifying healthcare beneficiaries and their demographics. (2) More targeted health surveys and questionnaires such as the biannual DOD cross-sectional survey to health-related behaviors. (3) Manage demand to identify potential problems early and intervene quickly, a "right place," "right person," "right time," before smaller problems became bigger ones. (4) Managing capacity becomes easier and more predictable with better knowledge of preventive and treatment needs of the current community. (5) In the optimized clinical system, evidence-based primary prevention—general education or training for the entire population—is best done within the community, supported by healthcare providers; secondary prevention identifies segments of the population and individuals at higher risk for disease/adverse outcomes—and targeted interventions are provided—that may include physical fitness training, nutrition, anger/stress/time management, spirituality, and nicotine cessation. Within this population, there will be individuals who meet a diagnosis for a medical condition for which more definitive treatment is needed. (6) With community outreach, the MHS is tied to the community, and with the right design for data systems to support this, the MHS is able to measure changes in health status and performance at both the individual and community level.

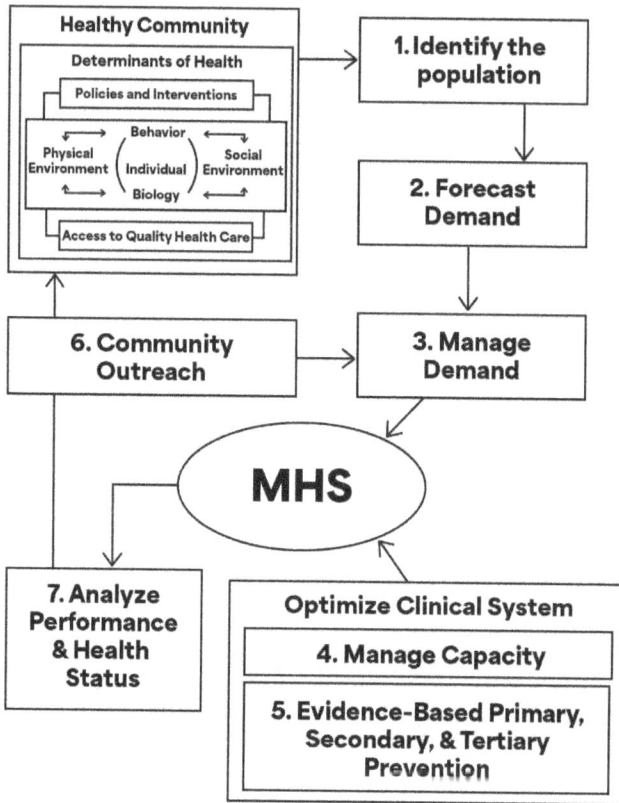

Figure 1.1

From the clinical view of a physician and the technological view of an IT developer, this book is for:

1. Direct-care clinicians to define what is necessary for documentation that improves the quality of the patient encounter, improves clinical decision-making, and facilitates rapid assessment of patient progress—and to better articulate the functional requirements for a patient-centric EHR.

2. Healthcare systems administrators to understand how EHRs collecting proper data will also serve all business needs and reporting requirements: improving access to care, better care, collaborative care, coordination of care, and greater provider satisfaction.

3. Insurers and government healthcare systems to understand how EHRs can support transformations to a population health model that lowers overall costs of healthcare, eliminating activities that add no value to positive outcomes and identifying interventions that increase morbidity and mortality of their covered beneficiary population.

4. Healthcare researchers to recognize how an EHR optimized for patient care provides data necessary to assess the impact of real-world short- and long-term interventions within the entire population of patients beyond the scope of randomized clinical trials.

5. All individuals to understand how EHRs today have interfered with the patient encounter because they contain incomplete, fragmented, redundant, conflicting, or incorrect data—and to advocate for an EHR that improves the clinical encounter with data necessary for shared and informed decision-making, with less clinical documentation time. We believe this will result in endless opportunities to make potential improvement from every encounter in care that leads to better outcomes by supporting technology that truly supports a clinical encounter.

Throughout this book my coauthor, Gilbert Pant, and I will be discussing the missed opportunities and the potential of EHRs. Given Gilbert's expertise, he will be dissecting the problems from a technological perspective, while I will be primarily focused on the clinical perspective. *Connecting Care* is organized in a way that represents these two viewpoints and it is our hope that readers will consider them both equally.

In chapters that follow, we address and discuss:

- What was good: an early emphasis on healthcare intended to improve health and why many went into medicine—with evolving technology that truly improved the quality of care provided.
- What went wrong: with a hardening emphasis on facility-based care and focus on profit—as providers develop burnout and moral injury—many now leave the profession or abandoning direct patient care.
- How this happened: the emergence, integration, and dominance of Practice Management and Scheduling Software under the EHR label—EHRs shifted from "help the patient" to "grow the business"— using technology to maximize justifiable reimbursement.
- How we fix it: "Back to basics" using new and updated technology refocused to help take better care of patients while serving various business needs in the background. Business grows because of better patient care and new knowledge rapidly integrated into best practices.
- The underlying premise: Design an EHR to help a clinician take care of a patient, and the resultant data collected will be of the quality necessary to meet the needs of administrators, insurers, and researchers that improves access to care, improves patient outcomes, and lowers healthcare costs.

PART I

The Early Evolution of the EHR

CHAPTER 1

By Ken Hoffman

A Clinical Perspective

Clinical Care and Documentation

Good clinical care requires accurate and concurrent documentation.

The Early Intersection between Paper Medical Records and Computerized Records

In the dawn of computers and electronic patient records, a primary purpose of paper documentation had been to enhance communication and collaboration between a patient and provider and between different providers who all might be involved in the care of one patient. A comprehensive history included a timeline of various problems and interventions and the impact of prior interventions. Included were other key and critical aspects of medical history, such as vaccinations, medications, allergies, hospitalizations, surgical procedures, and specialty consults.

Within each episode of care, every planned documentation has been the means through which each provider could remind oneself of patient health status, medical/psychological/social problems that had become a current focus of attention, treatment plans that guided interventions, monitored patient progress, whether desired outcomes had been achieved, or whether there were complications achieving desired outcomes.

Clinicians had the power to create their own patient or client records, which was believed to be necessary for better patient care and could be designed to be consistent with the guidelines of the provider's specialty organization. These records could be used as needed by other clinicians, in peer reviews, for administrative coding, and when needed by other providers. A patient record was intended to document the work done, diagnostic and problem evaluations, treatment plans, and decisions made.

An Early Computerized Provider Order Entry (CPOE) System: The "Light Pen"

In 1980, I was exposed to an ingenious use of technology. Through a computer terminal that had a light pen and keyboard, I could access a list of my patients to enter and send orders and retrieve results. The "Light Pen" system was easy to use. We only needed fifteen minutes to learn how to access the system with a unique password. It was intuitively obvious. I simply clicked on a patient to order or retrieve information. I could create or use a predefined order set for admissions that duplicated what I would have done on paper but faster and more legible. For orders, I could develop a standard order set—most useful for admissions and discharges—or order individually specific labs or procedures. If an emergency/urgent order was needed, I could Light Pen click on the "Stat" box. If the "Stat" box was checked, people came running to either collect the lab or start the procedure.

Results would always print to the printer at the nurse's station for inclusion in the medical record. If the "Stat" box had been clicked, I would be immediately notified. I could look up results at any terminal in the hospital. In looking up results, I could also Light Pen how I wanted them displayed, allowing for flowcharts on demand and easily seeing any critical changes in labs over time and by type.

The EHRs I've been required to use have not been designed as well as the Light Pen system for intuitive use that also quickly provided actionable data displays.

The First Patient-Centered Problem-Focused EHR: MARI and Theresa

At the time I was enchanted with the Light Pen CPOE, Henry Camp was developing a patient-centered problem-focused state-of-the-art EHR in Atlanta.

In the 1970s, the National Library of Medicine (NLM) had an interest in the aggregate analysis of medical records. Henry Camp, who had IT and programming expertise, connected with the School of Medicine at Emory University and the School of Information and Computer Science at the Georgia Institute of Technology, successfully applying for the NLM grant. The research components effectively used biomedical knowledge accumulating in many forms and within patient records. To accomplish this, Henry created the Medical Aggregate Record Inquiry (MARI) system that relied on accurate information within the patient record, and an existing highly formatted paper record system designed by physicians to provide better patient care.

Henry realized the only way for an accurate aggregate record analysis in MARI was to use technology to improve the quality of the patient encounter, helping doctors take care of patients. The EHR developed for this purpose, Theresa, operated for over twenty-five years, from 1983 to 2008, with 99.999 percent "up time," no security breaches, and less than one second retrieval time on any patient. With many patients in the system at the time, it was replaced by one of the large EHRs today. Fixing the problem today takes both clinicians and IT experts with a shared vision of healthcare that focuses on improving the quality of encounters between patients and providers—a similar concept to the vision that was behind Theresa.

Henry worked directly with clinicians for whom he designed Theresa, using IT to improve the quality of the clinical encounter. He intentionally designed the system to be "failsafe," with the vision that a down system is a dead patient. He treated Protected Health Information (PHI) as "Top Secret," designing information storage and transmission to prevent hacking and unauthorized access. Theresa was designed to interface with any medical system for continuity of care and compatibility with other systems.

Paper records became electronic records that could be mined for medical outcomes. Camp had the insight to realize that accomplishing this would require understanding the needs of a physician within the time allocated for the clinical encounter: collect only necessary clinical data relating to diagnosis, diagnostic tests, drug therapies, etc., and then provide information on the clinical decision-making processes clinicians used when treating patients. The only way he saw to make this possible was to be an observer during the treating physician's care of their patients.

Screen designs and screen flows were created to efficiently and accurately document a clinical exam. Data became immediately available for ad hoc natural-language queries. "Theresa" was compatible with any hardware from any generation—from VT100 terminals to the latest Apple product—and inexpensive to maintain relative to other EHRs at the time or today.

Theresa is a patient-centered problem-focused EHR, developed for clinicians seeing patients and tracking patient health status and treatment over a lifetime.

Independently, our IDEF work led us in the same direction.

The Army Alcohol and Drug Prevention and Treatment Program: Optimizing Clinical Work

Embedded in population health improvement is the need for evidence-based medical care. By design and through policy and leadership support, the Army's alcohol treatment programs had developed a systematic approach across all clinics worldwide for comprehensive bio-psycho-social assessments and treatment plans for which there was consistent paper records maintained at each clinic, kept at the quality needed to pass Joint Commission accreditation. These were kept separate from the EHR which did not have the security needed to protect patient privacy and confidentiality required under Title 42 Code of Federal Regulations part 2. The paper records were well organized and were evidence of the work being done by clinicians with every patient seen.

Using IDEF activity modeling methodology, in our study, the clinicians all agreed there were three equally important and exclusive activities related to patient outcomes.

Figure 1.0 shows the top-level view of a specific business process reengineering methodology: Integrated Definition Modeling (IDEF) that was used to define the work occurring in clinical care, resources needed, inputs that would be changed to outcomes as a result of that activity, what regulated the performance of that activity, and the activity-based costs.

The mission for all clinics: assess and treat all eligible beneficiaries identified with an alcohol or drug problem. The expectation was that referred individuals would be assessed for possible treatment, and if treatment was needed, this would lead to full remission (e.g., "rehabilitated individuals"). It was also of equal importance that novice staff would receive the education and skills needed to become "trained staff" to provide treatment a patient needed to reach the optimal outcome. Related to patient care, the clinic would receive multiple records and requests for information that would need to be processed and reported out.

From a "top-down" view with highest-level leadership support: Directives, Standards, and Guidance were all policies in law, DOD Directives/Instructions, Army regulations, professional practice guidelines (such as the American Society of Addiction Medicine's six-dimensional patient assessment), the Department of Veteran Affairs (VA)/ DOD clinical practice guidelines, and clinic standard operating procedures that would meet Joint Commission standards for accreditation.

A powerful justification for getting resources and time needed to operate the clinic always was applicable directives, standards, and higher-level guidance. Leadership support was unified at DOD—the deputy assistant secretary for health affairs, clinical programs, and policy, Dr. John Mazzuchi; the Army Surgeon General's addiction medicine consultant, Dr. Terry Schultz; the alcohol and drug program manager at Army Medical Command, Dr. Wanda Kuhr; and the quality program manager responsible for guidance and forms clinics were using to meet accreditation requirements, Dr. Chuck Deal. Related to medical records, I also found critical support from Terry Foley, who was responsible for meeting Office of Management and Budget requirements needed to have a medical form that could be used in a military medical record.

IDEF methodology also had "top-down" support. Throughout the 1980s and 1990s, IDEF methodology has been used to develop the Military Health Care System Enterprise Architecture. While the work we did was for a specific program, it was highly generalizable to behavioral health, general medicine, operational medicine, and population health improvement. Had I not used that methodology, I would not be writing this book, nor would I have the insight today of what's gone wrong with EHRs and how this might be fixed.

From a "bottom-up" view, we needed a very granular understanding of the work occurring in a forty-hour workweek for every clinician, administrative staff, and supervisor working in the clinic. This came from both talking with each individual and observing the work they did, including interactions with others. With this detail, an activity-based cost model could be developed along with the opportunity to identify work that added value for the desired outcome. For this detail and eventual full consensus, our IDEF modelers, Hudson Keithley and Marion Huck, were on-site with the clinical directors and staff at clinics in Fort Benning, Stewart, Gordon, Eustis, Knox, Bliss, Schofield Barracks, and West Point. Other subject matter experts from Tripler and Walter Reed Army Medical Centers, National Naval Medical Center, also volunteered significant amounts of time and information. With their input, an ideal activity model was derived and presented to the 104 participants attending the 1993 Clinical Directors Conference, San Antonio.

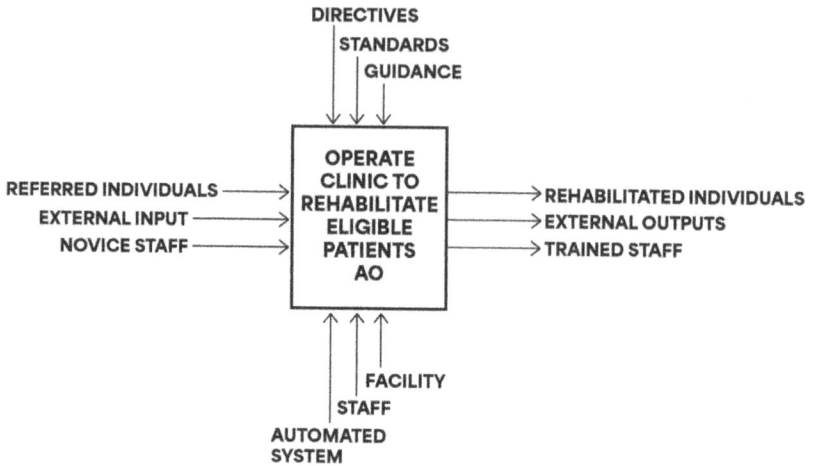

Figure 1.0

From the ideal activity model illustrated later in chapter 1, two programmers, Todd Neven and Steve Tanner, worked with Hudson Keithley to develop the data model that would be validated by clinicians in direct patient care before creating a proof-of-concept prototype EHR. The clinicians loved the result.

Before creating an EHR, the most important task was to capture the mission-essential work for each alcohol and drug treatment clinic throughout the Army worldwide. All would receive patients, either self-referred or through referrals from other sources. Each clinic was expected to follow established guidelines for triage, comprehensive biopsychosocial assessment, treatment planning, working with the patient for concurrence with the treatment plan, and enrolling the patient into treatment. Treatment also followed established guidelines that covered psychological and medical problems, which integrated the impact the social and physical environment would have on achieving treatment goals and objectives. Part of a comprehensive treatment plan included referral to other specialists when needed, with collaboration and coordination with those specialists. Part of the continuum of care included outpatient follow-up (or aftercare) for patients discharged from residential treatment. In these instances, the clinic would provide ongoing outpatient care for the following year. The clinic's connection with higher levels of care, other clinics providing the same level of care at different locations, and connection to community support programs provided a continuum of care and interventions needed for a patient to achieve and maintain long-term recovery.

Of equal importance, clinicians have limited time with patients, perhaps up to sixty minutes for a new or therapy patient and ten to fifteen minutes for an established patient. Within this time frame, providers must also complete all necessary documentation. What needs to happen during this time?

- Define the diagnosis or problems that will be a focus for treatment.

- Identify changes in the problem since last seen.

- Establish a treatment plan and how it can be implemented.

- Document efficiently to establish a clinical baseline, track progress, and for communication.

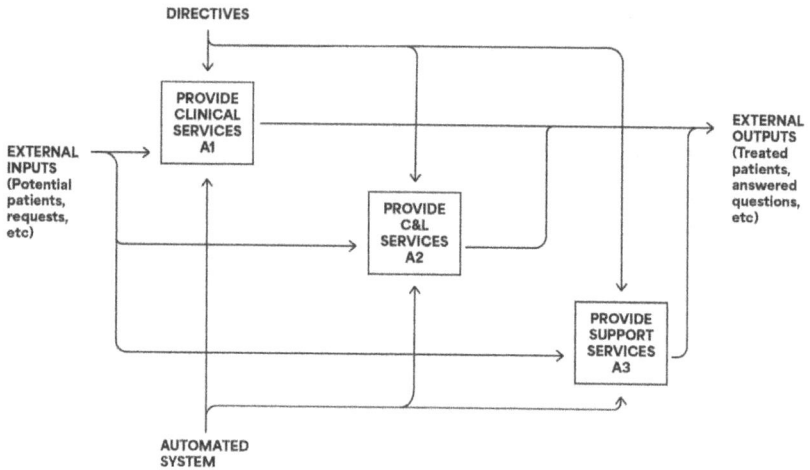

Figure 1.2

In population health improvement, more emphasis had to be given to consultation and liaison activities, to other organizations and leadership in the community, and with other providers in the same and other facilities. Primary prevention included providing general education and outreach about alcohol- and drug-related risks and general preventive skills such as time and stress management—skills that were also provided to patients in treatment. This was acknowledged with the significant time that was needed in providing support services within the clinic for training and supervising staff, justifying, developing and managing the budget, and assuring clinicians met both continuing education and licensing requirements to maintain credentials and improve service delivery. The automated system to clinical services would be the EHR, but there were other automated systems that had critical functions that were not related to medical health records.

Specific to providing clinical service: the automated system was to be our prototype EHR designed and focused to help the clinician provide the services necessary for the patient.

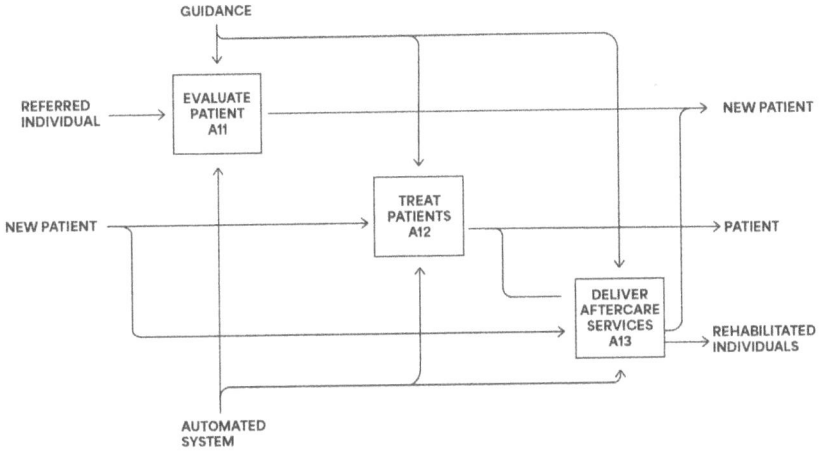

Figure 1.3

The important improvement from normal care at the time was recognition of the importance of patient evaluation that included triage, a comprehensive assessment, problem identification for treatment, development of a treatment plan, and with shared and informed decision-making, obtaining concurrence with the patient for the treatment plan. It was at this point a referred individual would be considered a "new patient" ready to engage in the treatment. Other referred individuals might have problems not appropriate or adequately treated through resources at the clinic, better referred to a higher level of care or to another clinical service. If appropriate and enrolled for treatment at the clinic, one component of treatment was referral to other providers and case management for problems being treated outside the clinic.

On paper, the Army's treatment counselors would document which signs and symptoms led to the diagnosis and specifically considered in developing the treatment plan.

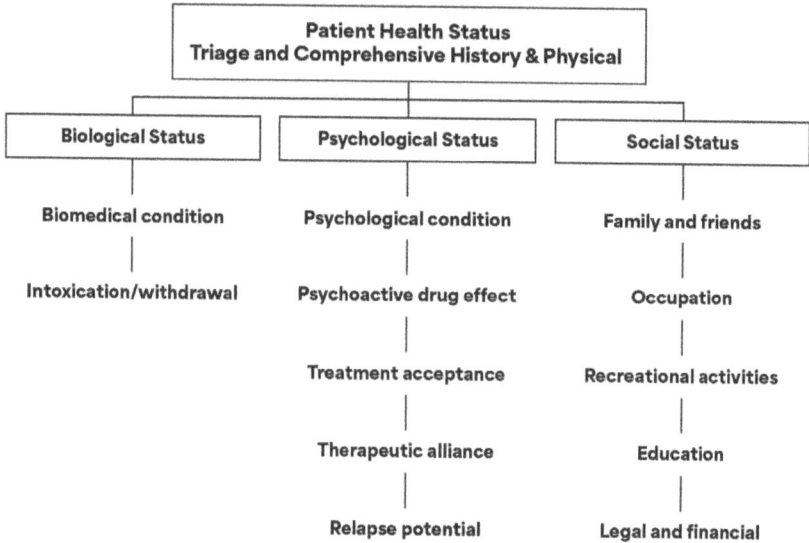

Figure 1.4

The paper record was designed to support the treatment guidelines and monitor changes in the patient's condition over time in treatment. As treatment goals were met, the patient would be discharged from clinic services but monitored through aftercare to periodically check if treatment outcomes had been sustained or needed follow-up care.

The major problem we identified was the amount of time—averaging about 50 percent of available clinician time—for documentation and reporting. Each new patient was generating thirty-one pages and fourteen forms for the medical record, while the clinic had ninety-one reporting requirements over the course of a year. As we were designing the prototype, I had received word that we had to have the fourteen forms used for patient intakes all in the EHR.

These forms had been developed over time with subject matter experts, and they were considered necessary as part of the patient assessment and included in the medical record as part of Joint Commission accreditation. They were completed during the initial assessment of the patient and then stored in the medical record. Once the assessment had been complete and the integrated summary was written, there was no further use, but it had to be retained for the medical record.

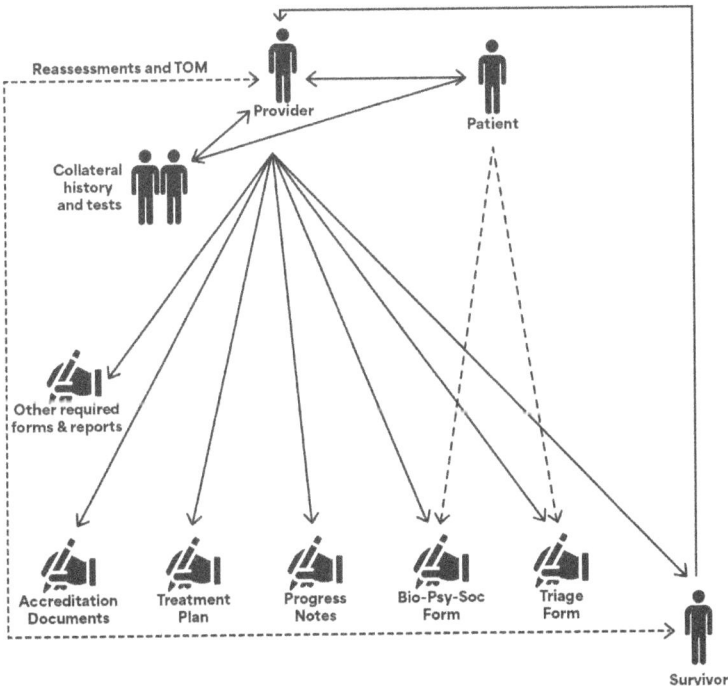

Form-based documentation: deceptively simple but time-intensive—consistent in all medical records in the worldwide network of Army clinics at the time. (Figure 1.5)

So much relies on the provider to record everything into the patient record, with the surveyor/auditor also required to read through charts and question providers to get a sense of clinic quality or detect problems. The patient may fill out forms that the clinician reviews, but it potentially could contribute far more.

Optimizing Documentation for Collaborative Treatment and Improved Outcomes

Using the optimized activity models—and knowing that the prototype EHR had to be fast, capturing the input-output data within the activity model and programmed using a relational database—Oracle, the programmers, developed a schematic, illustrated in Figure 1.6.

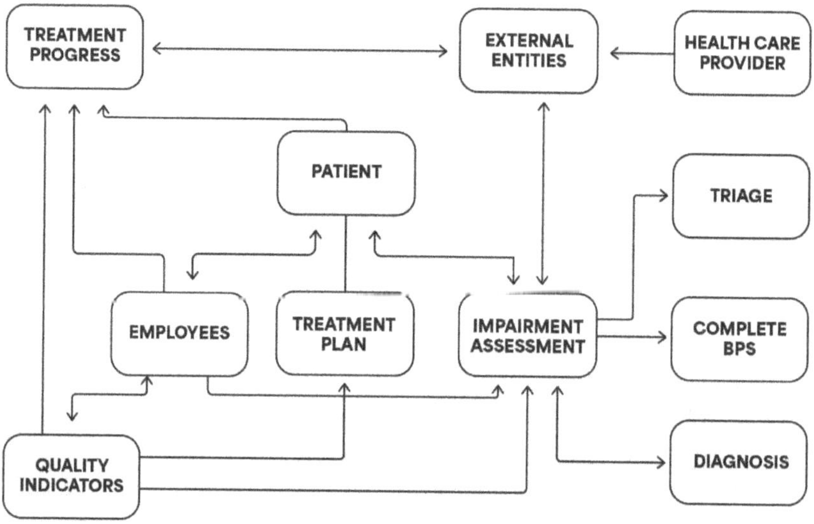

Figure 1.6

In the schematic, the credentialed counselors assessing the patient, developing the treatment plan, and treating patients were employees while physicians and other providers served as consultants, working in other clinics and sometimes treating the same patient for other problems. Together in the process of triage, doing a comprehensive bio-psychosocial (BPS) assessment, and making a diagnosis, it was the impairment assessment that drove the treatment plan and against which treatment progress was monitored. Over time, through quality indicators, it would be possible to better match employees with specific types of patients and problems for better outcomes. Prior to any programming, the programmers worked with clinicians through patient scenarios, building into the screen designs and screen flows, while the clinical guidelines supported the clinician within the time allocated for a new and established patient encounter.

The Conceptual Change: From Form-Based to Patient-Centered

When programming was completed, the clinicians loved the prototype EHR! We estimated that documentation and reporting time could be cut to less than 10 percent by collecting only necessary data without redundancy and using only that data to generate any required report without further clinical time.

Patient-focused documentation with "one-write" input allowed data collection to be distributed between provider, patient, and collateral sources. From inputs, reports can be efficiently generated and given to the "right" person, at the "right" time, in the "right" format, and the "right" amount.

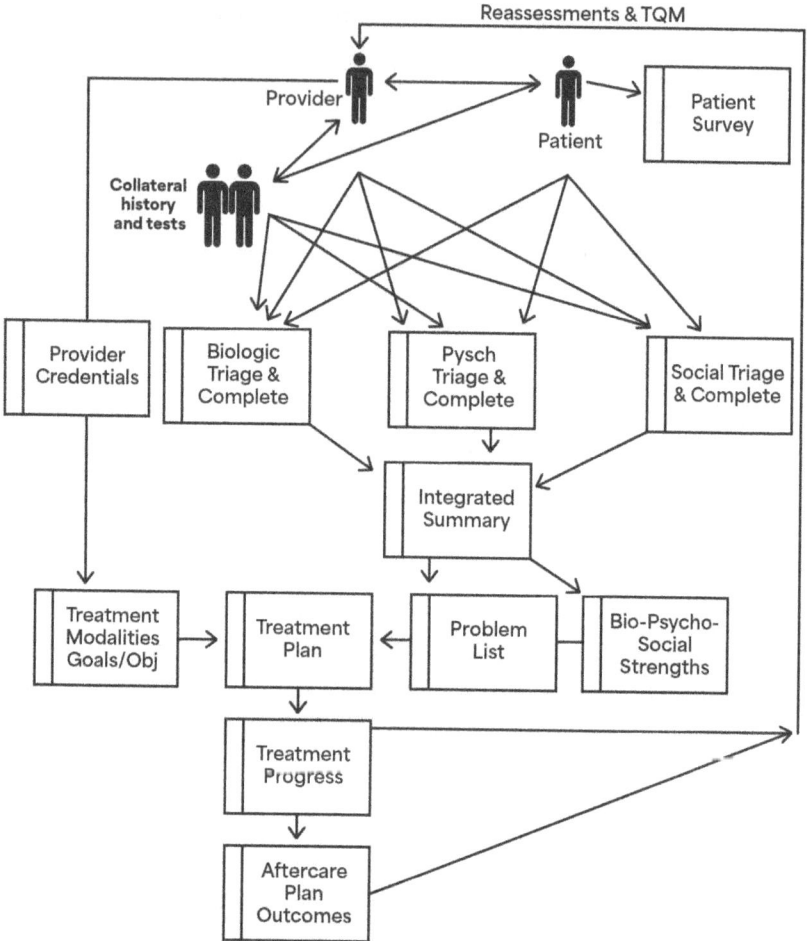

Figure 1.7

The notional model provided the opportunity for distributed data entry by individuals most interested in entering essential and accurate data that will be immediately useful.

The most critical part of the EHR was the integrated summary. This included both specific problems and their severity, along with identified strengths that could mitigate the level of care or intensity of treatment. The problem list could then be matched to specific treatment modalities recommended for treating those problems. Connected to treatment were specific goals and objectives.

We realized that simply having a ICD or Diagnostics and Statistics Manual of Mental Disorders (DSM) diagnosis was insufficient and inadequate. It was equally inadequate to define treatment in terms of Current Procedural Terminology (CPT), Diagnostic Related Group (DRG), or Healthcare Common Procedure Coding System (HCPCS) codes. These were closely tied to reimbursement without the specificity needed to know more about the specific problems, treatment modalities, or treatment goals and objectives.

With the design of the prototype EHR, we saw the path to track health status changes as a function of treatment progress reaching defined goals and objectives and outcomes management, leading to more personalized treatment plans and outcomes management.

Multiple methodologies (self-report, proxy, clinical exam, laboratory) and information sources allowed assessment cross-validation to arrive at "truth."

The Patient Portal Solution to Intensive Clinical Documentation Requirements

Given all enrolled patients had to have a documented substance use disorder, the critical issue for the integrated summary was which of the eleven signs and symptoms did the patient have leading to the diagnosis; the signs and symptoms were the focus for defining treatment plans and outcomes. Considerable additional time was needed to complete the fourteen forms that were to be part of the medical record.

To decrease the documentation workload and time for clinicians, we came up with a solution: let the patient fill out the forms in a patient portal. When interviewing the patient, the clinician could review the forms in the EHR completed by the patient and acknowledge they had been reviewed. The endpoint was to guide the development of the problems in the integrated summary. If the forms were not used, there was the option to use a comment box for the rationale for problems in the summary.

Decision-Support Software

Whether documenting a clinical encounter on paper or in an EHR, there was a need to have up-to-date knowledge for immediate use, integral to the decisions leading to a diagnosis, the need for additional information, and the shared decision-making with patients for treatment recommendations and alternatives.

A More Clinically Relevant Use of a Patient Portal—Instant Medical History (IMH)

Instant Medical History (IMH)[2] represented a breakthrough use for a patient portal. IMH started in 1989 by Allen Wenner, a family practice physician, when he realized much time could be saved if the patient directly provided data needed for the patient visit. IMH was patient-interview software that organized patient concerns and essentially provided the subjective part of a medical history for the physician.

The patient would answer a series of yes or no multiple-choice questions in a kiosk. The internal algorithm selected follow-up questions based on the prior ones answered. The software would translate the responses into clinical terminology and organize positive and negative responses by the relevant organ system, generate a

---

2   Wenner A. R., Ferrante M., and Belser D., "Instant Medical History," *Proc Annu Symp Comput Appl Med Care* (1994): 1036.

medical record report (below), and export the results to the EHR. The patients didn't mind using the computer, and the algorithm could capture more than one problem or prevention needs that could all be accomplished within one visit. The time needed for documentation was greatly reduced, with the subjective aspects of history coming directly from the patient with IMH and the IMH report imported into the EHR. Dr. Wenner found it possible to see more patients per hour with more accomplished in a patient visit, and sometimes without need to refer out for care possible in a primary care office.

History of Present Illness

#1. "Fever":
Duration: He reported: Fever three to four days.
Timing: He reported: Fever in the last seventy-two hours and abrupt onset. Highest fever within past forty-eight hours. Illness began in foreign country.
Severity: He reported: Fever decreasing daily. Highest fever 104F or 40C
Context: He reported: Night sweats.
Associated Signs and Symptoms: He reported: Fatigue. Pain associated with fever. Fever associated with rigor and night sweats.

IMH is still available today. It is used by over a million patients daily in ambulatory care clinics. It is integrated into most commercial EHRs. Today, IMH is not used primarily for improving quality of care by giving the doctor all the information to make the correct diagnosis. Health system buyers of IMHs want only their disease-specific EHR template filled in to ensure maximum reimbursement. Since the underlying EHR database contains a limited number of defined subjective fields, IMH is truncated to collect only that information.

Epocrates®

In the early 2000s, I discovered a remarkable software product, Epocrates®, that justified my purchase of a PalmPilot®.

In a very small operating environment, Epocrates was clinically intuitive, easy, and quick. It had several modules—a rapid drug reference that would automatically update when synchronized on a computer. I could search for any drug by name or class, with the initial screen displaying adult/pediatric dosing and covered contraindications and cautions, drug interactions, adverse reactions, cost and package information, and other information. This even included pregnancy drug category, lactation recommendations, metabolism, excretion, Drug Enforcement Administration (DEA) schedule, and mechanism of action. There was a notes page that could be used for writing personal notes about the medication.

There were other modules for tables that included advanced cardiac life support (ACLS) information, inhaler colors, narcotic equivalents, therapeutic drug levels, pediatric immunization schedule, and thrombolytic criteria. A drug interaction module allowed the construction of a drug list to test for drug–drug interactions.

It included/interacted with a short "5-minute consult" that briefly described diagnostic criteria and treatment recommendations for a variety of diseases, to include psychiatric conditions for which I was evaluating and treating patients. From "5-minute consult," there were hot links to the recommended medications in the Epocrates database and treatment guidelines that had been published by the *British Medical Journal*.

In the thirty seconds I might have to prepare for the next patient in the waiting room, I could cross-check my potential options for the clinical encounter with Epocrates and use it with the patient for the fifteen minutes we had to discuss treatment and medication options.

Outside the office, Epocrates also had "DocAlerts" to provide up-to-date/urgent health warnings related to changes of drug dosing or drugs removed from the market.

There was a "Continuing Medical Education (CME)" module for continuing education credit. Since I commuted daily by metro, where there was no network connection, I was able to accumulate over 650 hours of CME credit.

At that time, there was a more encyclopedic software product, Lexicomp®, that provided more depth than Epocrates®. It was slow and more difficult to navigate on the Palm, but in the computing environment today, it has become a "go-to" program for use in my clinical encounters, providing more pharmacological details than its companion program, Up-to-Date®, that has more detail on related clinical practice guideline.

"Dashboard" Graphic Data Displays of Complex Data: GIFIC® (Graphical Interface for Intensive Care)

CPOE has the ability to immediately display complex actionable data that allows quick assessment for immediate intervention when indicated—and was done in 1990. Michael Lesser, MD, a cardiologist, taking care of patients in an intensive care unit (ICU), developed a character-based display to allow for a quick and accurate assessment of the enormous amount of data generated by labs and monitors that require continual monitoring to detect changes quickly in the patient's health status.

Dr. Lesser created GIFIC®, that for the ICU was a homunculus display of a patient lying down in an ICU bed with an IV pole. This graphical display using basic colors in discrete blocks allowed a viewer to immediately detect critical problems (e.g., abnormal lab values or problem with an IV). Each block represented a critical data point related to labs, and other parameters, color coded to indicate whether the value was within normal range (green) or abnormal (red), with changes quickly interpreted on the display. Clicking on a block would provide the underlying data.

For a cardiologist, it was possible to walk through the ICU and immediately capture whether patients were getting better or developing more problems—and know what corrective actions were urgently needed.

Figure 1.8 shows what this model looked like, but in gray scale.

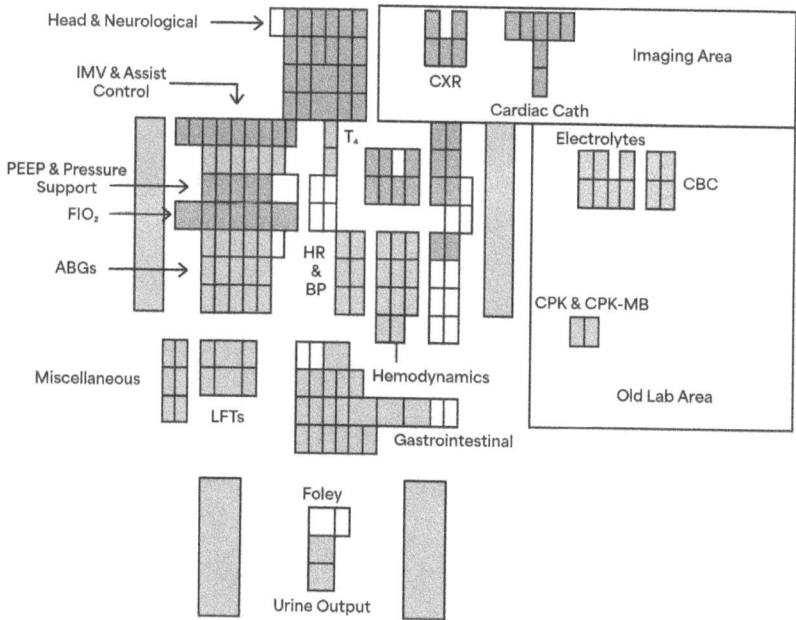

A gray scale model of GIFIC® (Figure 1.8)

I have yet to see any display similar to GIFIC, in any EHR I've used—and would love to see anything similar for mental health.

A Model Example of IT Support for Individual Health Improvement—the HRA

Between 1980 and 1995, a Health Risk Appraisal (HRA) had been developed specifically for the Army Health Promotion Program for use by the Army's Community Health Nurses (CHN). Independent of hospital medical services, health promotion was connected to command and garrison health promotion programs with two purposes: (1) Collect health data from individuals at the time nurses were available to interpret health-related risks and providing guidance for individuals taking the HRA. (2) Results could be used to report to each command the aggregate health status of soldiers. To protect the confidentiality of the soldier, HRA questions were answered on a computer card that could be scanned by card-feeder into the computer for interpretation. Following the printout of the interpretation, given to the soldier and used by CHN for advice and intervention, the unique soldier identifier was deleted. Aggregate reports on population health status could be provided to community and medical leaders, without identification of any individual who took the HRA.

The HRA was designed to enhance the interaction between a person and provider, providing evidence-based and prudent advice on any risk identified. It required no additional documentation from the provider.

With HRA data, it was possible to forecast demand for services and manage that demand with general education to the community and individualized early intervention for individuals at higher risk, referring those individuals who needed more comprehensive evaluation and treatment.

Traditionally, HRAs, when done independently of a patient encounter and as part of a community-based health promotion program, would not be included in an EHR. But, they would represent the collection of data that is processed to identify potential modifiable health risks that can lead to specific health-related interventions or referral for medical assessment/treatment.

Travax®

In preventive medicine, I learned to run a travel clinic. This required current knowledge of health risks and preventive measures that might be required for a traveler visiting any country outside the United States. Key references came from the Centers for Disease Control and Prevention (CDC), World Health Organization (WHO), US State Department, and other reliable sources for country-specific health information. Travax® was created to provide the exact knowledge needed and actions necessary within one clinical encounter.

All I needed from the traveler was the travel itinerary in the order of countries to be visited. The program created the report needed by the clinician and traveler that included required and recommended vaccinations, prophylactic medications, health alerts and concerns, and prudent advice regarding health-related support and safety risks within each country to be visited.

For example, a vaccination requirement or recommendation could exist for a vaccine not normally given in the United States, such as yellow fever, cholera, or rabies. A health risk in one country could create a vaccination requirement when visiting another country later in the itinerary. For many countries, there is required or recommended for malaria medication prophylaxis. Depending on the malaria parasite present, and the traveler, the specific prophylaxis could change. Related to US population health: a traveler not aware of risks or ignoring vaccination or prophylactic medications also could put others at risk for a disease no longer common in the US should the traveler become infected overseas.

This type of "just-in-time" education and decision support unquestionably improved the quality of clinical encounters I had with patients and fed my need for the most current relevant knowledge I needed to take better care of patients.

With GIFIC representing a visual dashboard of immediately understood parameters of data most easily collected in electronic systems and IMH reflecting a clever algorithm that patients completed prior to the actual clinical encounter—integrated as the subjective part of clinical EHR documentation—Epocrates® represented the next advancement for EHRs where this level of clinical support would be within the potential an EHR might have for improved clinical care.

When I used Epocrates® for decision support as part of the shared decision-making process, I would document that use in my clinical notes. This software made me feel good knowing I had reviewed and covered the most relevant and current knowledge for the discussion and decisions we were making.

Had a new disease been introduced into the US through returning travelers, "Doc Alerts" in Epocrates® could broadcast the public health risks and recommended action to all Epocrates® subscribers while any changed recommendations could be reflected in Travax® updates that today are near real time.

Clinical decision-support tools have a unique role: applying data and science, in a therapeutic patient relationship, to both enhance preventive health for patients without disease and to develop optimal treatment plans for clinically ill patients, thereby achieving optimal outcomes. A joy in medicine has been the opportunity to learn new things within each clinical encounter and apply that knowledge when working with patients.

Rather than this level of decision support, today I am numbed by the inane alerts I repeatedly receive that have little or no relevance to the clinical encounter. I now read those alerts, feeling the need to answer them simply to meet the needs of those who designed a program that created the alert—rarely, if ever, learning anything I did not already know before the alert.

Integrating a Lifelong Individual EHR with Individual Health Improvement

With the ICD providing a worldwide dictionary of specific signs and symptoms associated with each diagnosis and significant health-related problems, the foundation for a lifelong individual health record exists— where any clinician at any location may continue to work with the patient. That same record has the ability to merge health data from non-medical locations, such as wellness/fitness and employee health programs that lie outside the medical healthcare system.

Using the Army alcohol and drug program as an example, this was the vision:

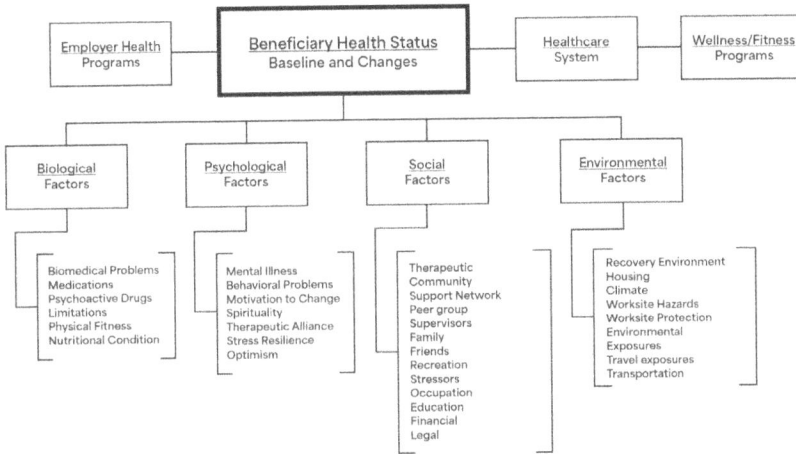

| Employer Health Programs | Beneficiary Health Status Baseline and Changes | Healthcare System | Wellness/Fitness Programs |

| Biological Factors | Psychological Factors | Social Factors | Environmental Factors |

| Biomedical Problems Medications Psychoactive Drugs Limitations Physical Fitness Nutritional Condition | Mental Illness Behavioral Problems Motivation to Change Spirituality Therapeutic Alliance Stress Resilience Optimism | Therapeutic Community Support Network Peer group Supervisors Family Friends Recreation Stressors Occupation Education Financial Legal | Recovery Environment Housing Climate Worksite Hazards Worksite Protection Environmental Exposures Travel exposures Transportation |

Figure 1.9

The individual is fully described accurately in data that captures a diagnosis and problem of clinical significance that become targets for assessment and treatment within various medical and behavioral health clinics from outpatient through inpatient care with the specialists best trained to assess and treat specific patient problems. At a healthier level, the same data becomes the foundation for interventions outside the healthcare system and covers the "social determinants of health" (SDOH) that generally have been outside the scope of traditional medicine and more reliant on community support services. This lifelong health record provides the collaborative relationship between other community services within a healthy community and healthcare provided within a medical treatment facility.

The Rise of Personal Health Records (PHRs) and Potential for EHR Interaction

Under full control of the individual, outside any clinical setting, a PHR can contain the same information as an EHR for continuity of care, with the same level of privacy, security, and confidentiality expected in an EHR. As an individual becomes interested in health-promoting changes, PHRs have been designed to help track health-related factors like exercise, diet, vital signs, sleep, and stress in real time for immediate feedback to the individual. For evaluating and managing medical problems, home monitoring devices can also be connected to a PHR that would be clinically invaluable if an EHR is designed to allow for the import of data for clinical evaluation and potential action. The health app today on my smartphone and smartwatch has far greater use and actionable detail than medical data being typed by my provider or technician into the EHR during any office visit.

Continuity of Care and the EHR

From the patient view, there is an expectation that the EHR has been built so medical data can flow with the patient to the providers.

Within the development of Health Level 7 (HL7) data standards, there is an exchange model for clinical documents: the Clinical Document Architecture (CDA) which has led to US specifications for a continuity of care document (CCD).

The CCD dataset represented a summary of the patient's health status to include problems, medications, allergies, and basic information about health insurance and care plans.

While I worked at Substance Abuse and Mental Health Services Administration (SAMHSA), I met Jim Kretz, who became a close colleague and friend with a shared vision for the EHR we needed. He had spent almost thirty years developing both clinical and healthcare administrative systems, including the first comprehensive EHR for Reproductive Endocrinology and Infertility Practices. At SAMHSA, he became the co-facilitator for the development of the HL7 Behavioral Health EHR functional profile and was co-chair of the HL7 Community-Based Collaborative of Care Work Group. There are worldwide data standards that would allow accurate transfer of health data needed by any practitioner who has permission to provide care to any patient.

Having used several EHRs over the past decades, to include changes in EHRs within the same clinic, I have not seen a seamless transfer of patient health information at the level of a CCD, except within the EHR of a specific company. However, the most critical medical information necessary for continuity of care has been defined.

Data Security and Protected Health Information (PHI)

Both Henry Camp and Jim Kretz, in the design of their EHRs, were most concerned about data security needed for PHI and the specificity necessary to define who had access to what type of health information. The concept behind patient privacy and confidentiality developed to prevent release of paper records to individuals needed to apply electronically. Should PHI be sent or disclosed outside the medical environment, individually identifiable health information (IIHI) that had been protected under the Health Insurance Portability and Accountability Act (HIPAA) or 42 CFR part 2 is forever lost. Unintentional disclosures and malicious attacks have become more frequent—harming patients in the process, that includes identity theft, sometimes at great expense.

CHAPTER 2

By Gilbert Pant

A Technological Perspective

On January 10, 1984, I was sitting in the waiting room of a hospital waiting for the birth of my son. The obstetrician joined me in the waiting room and asked me what I did for a living.

"I'm in computers," I replied.

"Ah, computers," said the doctor, "I hate computers. I've never met a single computer guy who understands what we do. The hospital's systems are just awful."

"I'd be very interested to see what you do," I said.

"No," he said, "there's no point. It will just be a waste of your time. Computers are a waste of time with hospitals."

"Tell me what it's not doing now that you want it to be able to do."

Given the green light, the doctor began listing for me all that was wrong with the system ending with, "… there is absolutely nothing on the screen that is remotely relevant to me or the care of my patients."

And so began my transition from building banking systems to healthcare IT. In the 1970s, I had worked on systems for banking. Successful computer systems were being used in academia and in certain areas such as the banking industry, and by the early 1980s, it had been implemented in several other industries, including airlines and major billing and payroll systems. These systems were called MIS (management information systems) and were intended to provide top management with a picture of what was going on.

The cost, at that time, to invest in a computer and in developing programs meant that only the largest organizations could afford a system, and so that's how the solutions were developed for.

By 1982, I had been coming up against barriers to further innovation in the banking field and was in search of a new challenge. This coincided with the development of the IBM PC, or home computer, which, although designed to be used in the home, had far-reaching consequences for large computers in main frames. The birth of my son opened the door to what would be my next innovative endeavor: Electronic Health Records.

Once I began to listen and observe the obstetrician and his staff as they went through their day and how they utilized the hospital's software system, it was immediately apparent that what the doctor, who as it turned out specialized in gynecologic oncology as well as obstetrics as an aside, had said was true: the information that he was able to access on the screen and what he was required to input was not helping him in his work nor improving the care he was providing to the patient. The system was set up for the general functioning of the hospital which did not correlate with the direct patient care provided by individual practitioners. Upon closer review, I realized he was working in a "workshop" with a group of about fifteen people whose needs, to improve care to his patients, from a computer system, were limited to his own workgroup which needed a system that enabled them to communicate and share data easily, just in their environment.

The concept of efficient, data-sharing workgroups was familiar to me. I had already been working with Christian Schumacher, the initial developer of the work structuring theory which defines small work groups of five to fifteen people as the most efficient and productive way to work.

Christian asked our team if we could design computers for work groups. This coincided with our exploration of PCs being used to provide software on a smaller scale for interlinked and intercommunicating smaller tasks, rather than the continued development of large-scale corporate systems. The first project we did was for one of Britain's largest industrial concerns, Imperial Chemical Industries (ICI), which manufactured everything from paint to chemicals. We went with Schumacher into one of their factories in Scotland and one in England and worked with him to split the organization up into small work groups. The idea was that each work group had its own inputs and outputs that they processed themselves, where the individual workgroups worked autonomously with other workgroups using interlocking but autonomous systems based on IBM PCs.

Each department (workgroup) treated the other departments in the organization as customers. We designed computer systems to support that concept. Each department would be told, "Here's your work group that is sitting in your computer, and you need to treat your workgroup as if it's your whole world, and everyone who is not in your workgroup is treated like a customer." This was immensely successful, and we worked with Schumacher for several years until he branched off into other avenues.

This laid the foundation for my work in healthcare IT, providing my team and I the knowledge and experience to design a system just for the obstetrician's work group. In other words, we recognized that a hospital is not one thing but a complex honeycomb of clinical workgroups all doing their own thing, using the skills and knowledge that they have and that only communicate with other workgroups when they need to, while communicating as much or more with other workgroups (in "competitive" organizations) engaged in the same type of activity for the exchange of data, knowledge, and technical help.

The traditional pyramid-shaped organization did not work within the clinical world, whereas it did exist in the managerial and administrative organization of the same establishment. It became obvious that a good EHR system needed to be dynamic and direct to any other group according to the need, and it should not allow management structures to be a barrier to providing the best available care while maximizing the efficiency of clinical staff.

It was through this experience that I came to understand that the whole approach to healthcare-based computing at the time was similar to the management information systems developed in the 1960s and 1970s in other application areas where the middle systems fed data up from the bottom to the people at the top.

It was clear to me that this type of structure didn't translate to the needs of healthcare for two primary reasons:

- Doctors don't care about the management information system; they only care about their patients.

- Every single small group of clinicians and every single small activity was separate from everything else that was going on. Yes, interaction did exist, but they were all doing different things. Different specialties like oncology, orthopedics, etc. didn't communicate with each other.

The result was that clinical support systems and patient administration systems might have some data overlaps, but they needed to be totally different systems with totally different design concepts.

This led to my development of the Template Health System in 1985. Template was the first system of its kind to formulate a system with these concepts:

- Healthcare is not one thing but rather a network of interacting systems.

- Medical data is always the answer to a question.

- Medical data, even subjective data, can be made possible by being coded.

- Clinicians can have a "view" of that data designed to fit their needs and work style which should be drawn from the data previously collected.

Over the next eleven years, Template was implemented into twenty-nine hospitals. Word spread of its success, and Reuters purchased the business in 1997. What Reuters didn't buy was use of my software outside of the UK. This led to my opportunity to work with ABT Consulting in Cambridge, Massachusetts, that same year. ABT Consulting was fascinated by Template Health System's ability to build large-scale patient information systems and asked me to advise them on building a database of mental health information for an organization called SAMHSA. That role eventually connected me with Colonel Kenneth Hoffman and our subsequent work in Korea.

PART II

The Korea Case Study

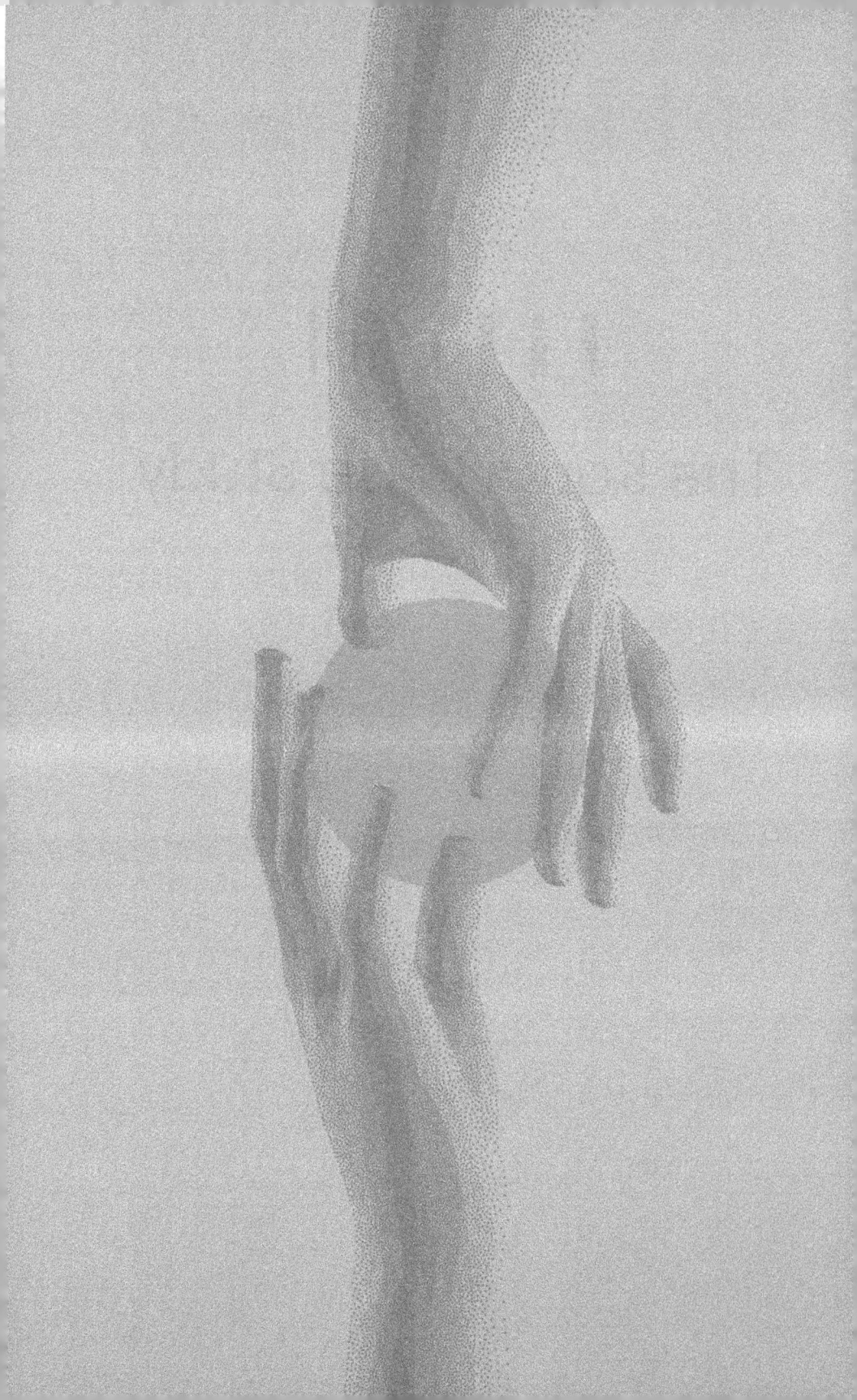

CHAPTER 3

By Ken Hoffman

A Clinical Perspective

Korea Case Study

In 1998, I was assigned to be the preventative medicine officer for the US Forces in Korea. Assigned to the 18th Medical Command, I was also the director of preventive services. Upon my arrival, there were three key problems to be solved concurrently: (1) developing a malaria control and prevention plan in coordination with the Korean plan, (2) monitoring implementation of mandatory anthrax vaccine immunizations for all US Forces, and (3) improving disease detection.

I had also received a $100,000 grant for an initiative in syndromic surveillance from DOD's Global Emerging Infections Surveillance (GEIS).

What was good: With individual initiative and common vision, the necessary technology has existed to achieve that vision, at a cost less than expected within short periods of time.

Developing a Malaria Control and Prevention Plan in Coordination with the Korean Plan

Malaria had become an emerging infectious disease, the same type of malaria that had once been in the United States but long eradicated. One risk of malaria in Korea was the potential to re-introduce it to the United States. While malaria is a leading cause of mortality and morbidity in many countries overseas, it is rare in the United States. To diagnose, a clinician needs to have a high index of suspicion—a high fever in many countries would be assumed to be malaria unless proven otherwise. The diagnostic test, a thick blood smear analyzed under a microscope, is not commonly done in the United States today, and treatment would be with the same medications used for prophylaxis.

Required for the spread of this malaria were the presence of a mosquito capable of carrying the malaria parasite and a human infected with the parasite. With a mosquito bite, the parasite from one infected person could be spread to a non-infected person.

The task was to identify which Americans should be required to take medications to prevent malaria infection, and for none who may have been infected to develop the disease.

This specific strain, Plasmodium vivax, had a unique quality: while some infected people would become ill within a short time, another group would become ill many months later with parasites that were dormant in the liver.

For any identified at risk for infection, there would be a requirement to take malaria prophylactic medications, one to cover the time the individual was in Korea during the mosquito season, and another medication designed to eliminate parasites that might be dormant in the liver.

I was quickly impressed with the focus of the Korean government and military on eradicating malaria, and how technology had been effectively leveraged to detect the mosquito vectors carrying malaria, identify individuals who had become ill, and implement preventive measures to halt the spread of malaria.

The Koreans already had the ability to map areas where there was known infection and areas at risk for infection both with population and environmental surveillance. I learned much from the Koreans as they were able to identify the point source of the infection and validate the strain through genetic mapping. Within a few years, the strain was eradicated in South Korea.

The importance for more comprehensive capability for syndromic surveillance became obvious as we prepared for the flu season.

Monitoring Implementation of Mandatory Anthrax Vaccine Immunizations for All US Forces

For all US Forces stationed in Korea, a mandatory anthrax vaccination immunization program was being put in place, with a required six shot series that would be documented three ways.

First, all shots by name were entered into a separate vaccination computer system tied to a Washington-based mainframe computer that was not linked to the existing EHR in use at the hospitals. The computer program was designed to send reports back to various levels of command, listing individuals who were overdue a vaccination by specific times. It also accommodated any exemptions to further vaccinations that had been determined by medical personnel. Additionally, all anthrax vaccinations were to be documented in the service member's paper medical record. And as a final step, all anthrax vaccinations were to be documented in the service member's yellow shot record, carried by the service member independently of the medical record.

While the vision was that this documentation process would be failsafe in documenting all who had been vaccinated, each method had significant problems. The computer system was slow, and generally data could not be entered within the time the service member was receiving the vaccination. This resulted in data entry outside vaccination times when service members had left the vaccination site or after hours—essentially humans doing key punch batch data entry at times when they could gain access to the computer. While there was an option to enter data by unit, it required checking the accuracy of names in that unit as to who was present and who was vaccinated.

The paper record presented its own problems. With large numbers of service members being vaccinated in a day, chart pulls for each patient were difficult. It was faster to generate a paper form that could be sent to the medical record room for filing. The paper records also had a defined structure and form numbers. Documentation for any one vaccination might appear in any one of several forms, or not at all if the medical paper record was no longer in the record room to which the forms were sent. Documentation in the yellow shot record relied on service members having that record with them at the vaccination site—a document that might be stored with the medical record.

From what I had learned in IDEF models and optimization, I thought it would be helpful to simplify documentation through designing a one-page medical note, that would be highly relevant by helping to improve interaction between soldier and provider, leading to the "correct" decision for vaccination or deferring vaccination, and identify any concerns related to the vaccination.

Within the alcohol and drug treatment program, we had developed a proof-of-concept rapid prototype of an EHR based on an optimized clinical model that minimized documentation time, collecting only data necessary for a patient encounter. We had been working from an optimized data model for this purpose, but the programmers knew the data model needed to be validated for the prototype. The programmers would create mock-up screen designs ("paper models") that could be displayed on a computer screen for clinicians to test. It was important to validate that the clinicians found the screen designs and flow helpful in the clinical encounter and in the shared and informed decisions needed between clinician and patient. Only when there was clinical concurrence with the screen designs and flow would the actual programming occur creating the prototype. Programmers explained the importance in cost: perhaps $1 to change a screen design, but $100 to make the same change once programming had occurred into the database. Clinicians were very engaged in the creation of an EHR they really liked.

It seemed obvious that a structured medical note for anthrax vaccination integrated directly into the clinical encounter at the time vaccinations were being given would be the optimal means to collect any concerns and adverse reactions about the vaccine. Vaccine reactions had been a strong concern among those who had developed medically undiagnosed symptoms. Medical records had not been designed to capture data and concerns when vaccinations were given.

For medical command, this presented a problem: an ad hoc fourth documentation added to three other documentation requirements wasn't going to happen.

Yet, a new physician, Cory Costello, assigned outside the medical command to a troop medical clinic, thought it an excellent idea and used the ad hoc form in the clinic where he worked. With all the other documentation requirements, the providers, servicemembers, and commanders liked this one-page form. It enhanced communication between the provider and servicemember, facilitating questions and answers, and was quickly completed with clear documentation as to whether the vaccine had been given.

Figure 3.1 is an illustration that is consistent with the approach of IMH where the patient completed the subjective part of the medical note.

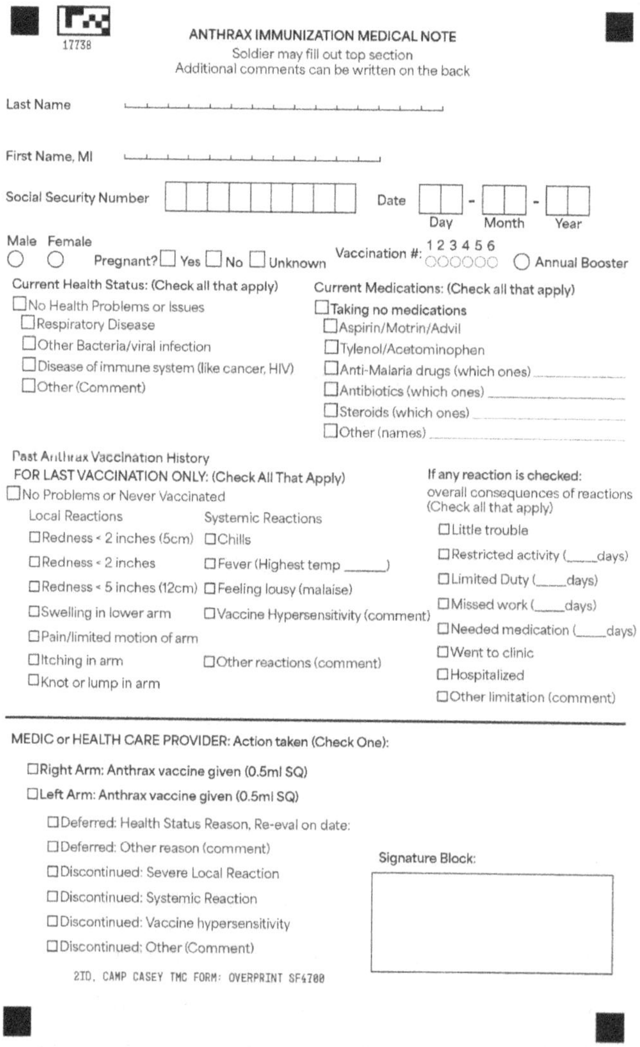

17738

ANTHRAX IMMUNIZATION MEDICAL NOTE
Soldier may fill out top section
Additional comments can be written on the back

Last Name

First Name, MI

Social Security Number

Date Day - Month - Year

Male Female

Pregnant? Yes No Unknown

Vaccination #: 1 2 3 4 5 6 Annual Booster

Current Health Status: (Check all that apply)
- No Health Problems or Issues
- Respiratory Disease
- Other Bacteria/viral infection
- Disease of immune system (like cancer, HIV)
- Other (Comment)

Current Medications: (Check all that apply)
- Taking no medications
- Aspirin/Motrin/Advil
- Tylenol/Acetominophen
- Anti-Malaria drugs (which ones)
- Antibiotics (which ones)
- Steroids (which ones)
- Other (names)

Past Anthrax Vaccination History
FOR LAST VACCINATION ONLY: (Check All That Apply)
- No Problems or Never Vaccinated

Local Reactions
- Redness < 2 inches (5cm)
- Redness < 2 inches
- Redness < 5 inches (12cm)
- Swelling in lower arm
- Pain/limited motion of arm
- Itching in arm
- Knot or lump in arm

Systemic Reactions
- Chills
- Fever (Highest temp ______)
- Feeling lousy (malaise)
- Vaccine Hypersensitivity (comment)
- Other reactions (comment)

If any reaction is checked:
overall consequences of reactions (Check all that apply)
- Little trouble
- Restricted activity (____days)
- Limited Duty (____days)
- Missed work (____days)
- Needed medication (____days)
- Went to clinic
- Hospitalized
- Other limitation (comment)

MEDIC or HEALTH CARE PROVIDER: Action taken (Check One):
- Right Arm: Anthrax vaccine given (0.5ml SQ)
- Left Arm: Anthrax vaccine given (0.5ml SQ)
- Deferred: Health Status Reason, Re-eval on date:
- Deferred: Other reason (comment)
- Discontinued: Severe Local Reaction
- Discontinued: Systemic Reaction
- Discontinued: Vaccine hypersensitivity
- Discontinued: Other (Comment)

Signature Block:

2ID, CAMP CASEY TMC FORM: OVERPRINT SF4700

Figure 3.1

This note was intentionally designed to:

- Collect only highly relevant information necessary to identify and discuss the patient's concerns, determine whether the vaccination should be given—and to which arm—or if deferring or discontinuing, the reason why.

- Summarize potential reactions and concerns that were in the vaccine package insert.

- Decrease provider documentation time with collaborative documentation—allow the soldier to fill out most of the form while waiting. The form is reviewed by the healthcare provider (HCP), enhancing communication between provider and soldier for answering questions or discussing concerns.

- Quick documentation by the provider of action taken and signed. In this note, the vaccine lot number was not included since all vaccines were from the same lot and being recorded into the electronic vaccine tracking system, and two other paper records.

- Designed as a "Teleform," it could be scanned into an SPSS statistical program database, allowing aggregate analysis and enhanced post-marketing surveillance.

The same note could be used directly for a medical decision to exempt the soldier from further immunization and as a report into the CDC Vaccine Adverse Event Reporting System. The provider would only need to complete the bottom half indicating whether the vaccine was administered and in which arm or whether the vaccine was to be deferred or had been discontinued.

The data was also analyzable directly from the note in aggregate. The patient completed the top half of the note. This highlighted potential concerns that would be part of the decision-making whether to vaccinate or defer for more complete evaluation. It also enhanced discussion between patient and provider about any concerns related to a prior vaccination.

From this note, in aggregate, I discovered from the first 400 soldiers vaccinated that there was a difference reporting side effects between men and women, and differences between those who were taking medications and those who were not. Continuing the analysis covering 2,800 vaccinations, these differences persisted throughout the vaccination series. In consultation with the Army's immunology consultant, this knowledge was immediately used to enhance health risk communication and provider training in a future vaccination. It also provided real-time knowledge that had not been possible before in prior vaccination programs.

Unknown to me, the Army's immunology consultant and Vaccine Health Center had used the same methodology to create their EHR, "I-TRAX"—and became a strong ally in the work we did. This type of note was unique enough; the consultant had the foresight to have this initiative covered under a Walter Reed quality improvement protocol.

Improving Disease Detection

In wartime, each Battalion Aid Station (BAS) would generate a daily Disease and Non-Battle Injury (DNBI) report that would go to the command and medical officer for early detection of health problems having impact on readiness.

In peacetime, no DNBI reports are done, with high reliance on lab results to signal problems of public health and command interest.

When arriving in Korea, a battalion surgeon, Cory Costello, MD, took the initiative to track illness for DNBI. Illustrated was the impact of tracking soldiers at sick call between September 14, 1997, through March 22, 1998, that covered the influenza season that year.

It was only when the prevalence exceeded 25 cases/1,000 soldiers did the hospital become aware of a respiratory problem in the field and reported it to preventive medicine (PM).

At the BAS level, there was awareness of 5 cases/1,000 soldier; earlier detection and greater knowledge of new cases over time.

Respiratory Diseases

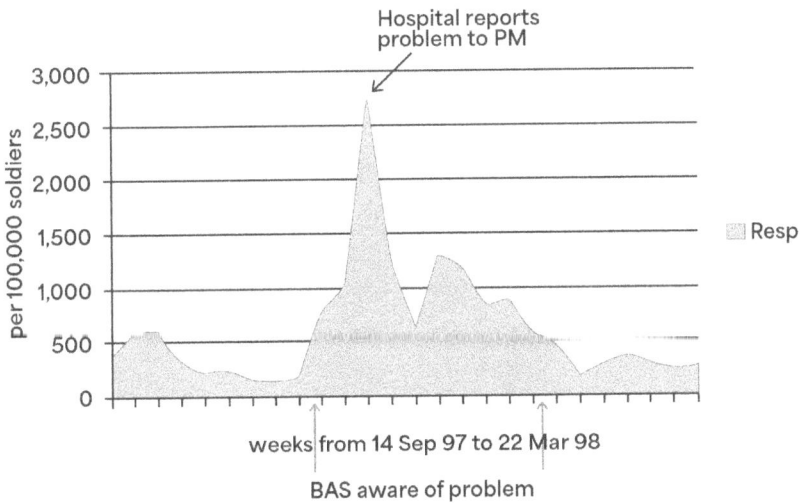

Figure 3.2

The development of EHRs designed for these purposes is still novel but achievable if the focus becomes truly patient-centered with the goal of helping the provider make the best assessments and treatment decisions based on best practices known at the time. An EHR designed with this goal also allows real-time changes that improve best practices based on patients being assessed and treated today by providers committed to helping the patient in the best way possible.

Using the triage medical note at a BAS for real-time syndromic surveillance: Three Completed Projects

With a heightened interest to improve detection of global emerging infectious diseases (GEIS), I received a pilot GEIS grant.

Taken together, these three projects were the foundation for the GEIS grant:

1. IDEF modeling in the alcohol and drug programs, then integrated into the DOD MHS Enterprise Architecture;

2. Ability to report quick results from the structured anthrax note;

3. Improved disease detection through DNBI reporting in peacetime.

With the purpose to develop near real-time surveillance for an emerging infectious disease, the goals were to:

1. Track disease rates at the BAS level without slowing down patient care nor creating additional work for the medic;

2. Provide reports to the preventive medicine officer that would identify:

   a) the most frequent sick-call problems that were not tracked (e.g., injuries, respiratory illnesses, gastrointestinal illnesses, dermatology problems); recognizing that many labs used for diagnosis were not done or not relevent at the BAS level;

   b) Health issues of high command interest:

   i. Infectious diseases (e.g., malaria, Korean Hemorrhagic Fever (KHF), sexually transmitted diseases, tuberculosis)

   ii. Environmental illnesses (e.g., heat and cold injuries).

   And that was how I first began working with Gilbert Pant.

CHAPTER 4

By Gilbert Pant

A Technological Perspective

In 1998, an opportunity arose for me to work on a project in Korea with US Army medical personnel.

The project brief was interesting and apparently quite simple.

In military operations, Disease and Non-Battle Injuries (DNBI) can quickly lead to miliary defeat. Under battlefield conditions, each battalion aid station (BAS) would be required to provide a daily DNBI paper report to command based on the symptoms reported by soldiers. At the BAS, each medic followed an Ambulatory Patient Care (APC) algorithm based on the patient's chief complaint to conduct a triage exam to determine if the patient could be treated at the BAS or needed higher level care. This would be documented in the paper medical record but was not in the hospital-based EMR. The hospital would become aware only of referred patients through labs sent from the BAS. Through the GEIS grant, there was an opportunity for real-time syndromic surveillance of DNBI conditions detected in the APC algorithms.

We could learn how many people were sick at any BAS. The

work was done in Korea with some development work in the United Kingdom and the United States of America, and the project was handled on the user side entirely by US Army medical staff, led by Colonel Hoffman.

The start of the project involved a detailed analysis of the processes in place, and the requirements of a potential system were identified and documented. There were a number of stages in developing a prototype to meet this need.

The first stage was to see when and how symptom data could be collected in a processable form.

The second stage consisted of how data could be centrally collected and consolidated.

The last stage was to see how the data, once consolidated, was to be used to identify clusters of illness outbreaks that signal an outbreak we wanted to control, and how quickly, so that appropriate action could be taken immediately.

The medics who first came into contact with a sick soldier in the BAS had emergency medical technician training. At a BAS, following medical and trauma protocols, they were trained to triage presenting signs and symptoms of a presenting illness or assess and stabilize traumatic injuries. Medics could follow published protocols, such as the Ambulatory Patient Care (APC) Model 21 algorithm which enabled medics to accurately identify and document specific symptoms and health indicators for triage and provide immediate care.

DNBI surveillance was essential for the early detection of potentially highly infectious diseases or conditions that would prevent deployment. Syndromic surveillance would detect signs and symptoms of conditions that could limit deployment or be early indicators of serious medical conditions.

So, we looked at how the medics collected this data in their recordkeeping and decided that a simple handheld device could assist the medics by making the recordkeeping more efficient by recording yes or no and numeric values as answers to questions.

These answers not only allowed medics to keep their records but also meant that the transfer of data between medics was simplified and removed issues that arose from poor communication caused by bad recordkeeping and poor handwriting. The handheld device required each question to be answered, often by a simple touch of a button.

The data we required centrally to monitor symptoms was therefore more likely to be accurate as it was collected as a by-product of the recordkeeping that a medic needed to do in any case, while providing him with a more efficient way of recording the data.

The next problem was to collect the data centrally, and this was done by transmitting the messages back to the base camp as an encrypted anonymized data string, simply providing the location, date, and time the data was collected.

Back at base, the data coming from the aid stations was collected and consolidated as soon as the messages were sent.

We had been provided with software by the UK Ministry of Defence that had been developed to identify population clusters of specific groups of symptoms, collected from the population at large. We adapted a version of this software to use here so that we could map clusters of symptoms against the relevant populations on a map, and the prototype proved that it would be possible to identify unusual clusters of symptoms, in specific locations or areas, that could mean an outbreak of infectious diseases.

And, as this was happening on the same day, clinical teams could be dispatched on the same day that symptoms were being reported to the places where there was the highest likelihood of an outbreak. This was significant as it meant that whereas previously one had to wait until patients developed further symptoms to conform their illness, with the result that diagnosis might be too late to save the patient, we could identify potential clusters quickly and dispatch teams to investigate and carry out testing to ensure that early detection was possible.

Once the proof of concept was complete, the findings were written and published. As well as providing an effective disease surveillance model, we improved the work of the medics in their day-to-day work, by collecting the data we needed for disease surveillance as a by-product of this improved method of work. The results were published, and we were encouraged to find a software company that would be prepared to develop a packaged system to be sold to the places where it might be useful.

We approached relevant organizations in the USA, Korea, and the United Kingdom.

The reaction was quite surprising.

In Korea, they saw the solution as useful for collecting data from general practice for a national disease surveillance program.

In the UK, the interest was limited to looking at it as a way of identifying low-level chemical warfare attacks on an unsuspecting public but was never implemented.

In the USA, there was no interest in developing a solution as each organization we presented it to thought the project was too small to be commercially of any interest.

PART III

The Chaos That Is

CHAPTER 5

By Ken Hoffman

A Clinical Perspective

- What went wrong: IT development was independent of the healthcare vision toward individual and population health improvement. IT also did not embrace the definition of an EHR but continued designing electronic medical records (EMRs) that did not support a population health improvement model.

- Both paper and EMR paper documentation became less clinically relevant for communication as they became more focused to meet various facility-centric business and administrative needs developed by contract and documentation for "defensive medicine."

Over time, the term EHR has been used interchangeably with EMR— with the assumption that EMRs could become EHRs. An EMR however is no more than the electronic version of a paper medical record. "EMRs are a digital version of the paper charts in the clinician's office. An EMR contains the medical and treatment history of the patients in one practice. EMRs have advantages over paper records. But the information in EMRs doesn't travel easily *out* of the practice. In fact, the patient's record might even have to be printed out and delivered by mail to specialists and other members of the care team. In that regard, EMRs are not much better than a paper record."[3]

EMRs have supported and strengthened facility-centric model of US healthcare.

The Administrative and Facility-Centric US Healthcare Model

Clinician productivity is measured by patients seen per day, and it justifies the workload through billing codes within a hospital or outpatient facility—both that can have different billing codes depending on the accounting system used.

A provider-patient interaction, or encounter, occurs in an inpatient or outpatient setting. They can be face-to-face or virtual through telehealth. Documents generated from each encounter will be stored either on paper or electronically to be used to justify medical necessity for billable and reimbursable events. Ignored in this model are non-patient activities that may have a significant impact on patient health, development of a clinical diagnosis, and prognosis for any treated condition.

Patient encounters will usually be in an outpatient setting, but there may be many outpatient settings, from primary to specialty care

---

3   Peter Garrett and Joshua Seidman, "EMR vs EHR—What Is the Difference?" Health IT Buzz, January 4, 2011, https://www.healthit.gov/buzz-blog/electronic-health-and-medical-records/emr-vs-ehr-difference/.

either within or through many facilities. At various times, there may be one or more medical conditions that rise to a need for inpatient care with inpatient facilities having various degrees of connectivity to outpatient providers who have been providing the care to the patient.

Each of the different outpatient providers and facilities will have a similar process that includes acquiring a patient for whom the facility/provider has the resources and competencies necessary to provide adequate care. The patient will have a designated provider who will develop a treatment plan to treat a specific patient condition, have a process for documentation, independently generate justifiable billing codes, and have a process for discharging the patient. This may be for various reasons including treatment goals reached, transition of care to another provider better able to treat current conditions, patient lost to follow-up, non-compliance with treatment plan, or change in insurance coverage.

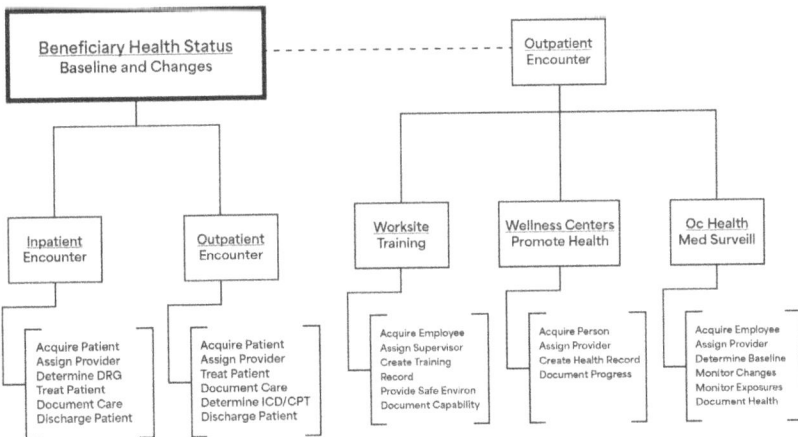

Figure 5.1

A similar process happens for inpatient care, but time allocated for this level tends to be short with high costs attached.

Completely outside the medical system are activities that the patient may engage in that have a direct impact on health improvement or in precipitating a new medical condition that requires medical or surgical intervention. Outside of inpatient or outpatient care, with 168 hours in a week, the patient may be having ongoing contact with various activities that either enhance health improvement—like health promotion and wellness programs, safety and occupational health programs, or worksite training related to physical and mental health—or diminish it.

The result of this siloed, non-patient-centered approach to individual health and related data collection is inefficiency, disjointedness, and redundant interventions and documentation, all of which result in incomplete healthcare: a model that inadequately supports an individual's health needs and leads to what we have today: the most expensive healthcare system in the world with life expectancy that is getting less over time.

Ideally, the military need for a "ht and healthy" force with healthy families—embodied in the concept of individual and population health improvement—would have led to an optimized person-centric health assessment and intervention model for healthcare. However, today, it conforms to the general American insurance reimbursement model, where an individual's medical records and other health information may be sequestered in many facilities outside the direct-care system and unavailable or unusable to any one provider within the time frame allocated for a patient encounter.

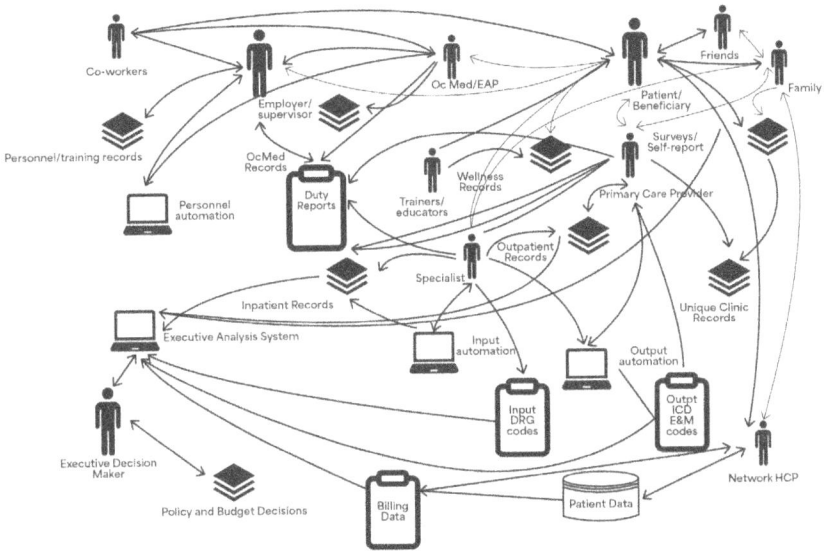

Figure 5.2

The beneficiary has the normal patient-provider relationship with a primary care provider and eventually with consulting and treating medical, surgical, and dental specialists. With the military health-care transformation, a greater reliance has been placed on referrals to TRICARE network HCPs who will have different medical documentation systems.

Related to health, the beneficiary has interactions with family, friends, fitness trainers and health educators, occupational health, addiction treatment and prevention specialists, employee assistance programs, co-workers, supervisors, and personnel departments: all of whom may keep separate paper and electronic records containing important health information specific to the individual's right and need to have individual health information. The beneficiary may also maintain a personal health record either on paper or electronically, taking advantage of health resources available to the general public.

Once there is a diagnosis made on any patient, that diagnosis begins to define the patient and potential healthcare needs. The same holds for prescribed medications where the medication itself could also be attached to a presumptive underlying diagnosis. What is more difficult to find in this model is the actual assessment and evaluation containing the diagnostic criteria that should have established the diagnosis. For a diagnosis that has a chronic or long-term component, it is challenging to find documentation for any points in time when the condition may have been in full or partial remission and more specific correlations with any medication or intervention that mitigated or worsened the severity of the condition.

From a cost and budget viewpoint, the system is fully reliant on billing codes that are provided independently of documentation of the clinical encounter, in which either a medical coder or the physician determines the billing code that reflects the work and time that is independently documented in either a paper or electronic record. In Figure 5.2, where the direct-care military healthcare system has contracted support for medical services, the medical records generated by various service providers are retained by the provider/facility, and billing codes are sent for reimbursement.

This has led to clinician treating patients in silos, feeling isolated, overwhelmed both with the need to document according to the implicit design of an EMR not designed to support the clinical interaction.

With so much technology and non-integrated software products used, direct-care providers are left with the full responsibility of seeing patients on time; reviewing past note(s) of varying structure, completeness, and accuracy related to the reason the patient is now being seen; and documenting with sufficient detail that will justify the bills the facility will send for reimbursement. Inherent in this process is the assumption that the provider seeing the patient has all the information necessary for the visit somewhere in the EMR and will put more information into the EMR related to the present encounter so that future providers will have access to the current encounter.

The critical and actionable patient-centric notes written by a clinician today will be buried beneath the information deemed essential for billing purposes. If continued care of a patient is necessary, the EMR records I read today make it difficult to determine if current prescribed medications and treatments have been helpful, if others have had adverse effects, or if there have been other components of treatment needed or others that have been detrimental. Also difficult is finding lab results that may be necessary or recommended based on the patient's condition and treatment, and determining whether they have been done. The EMRs I've used today have been designed to make it difficult to understand the thoughts or decisions made by other clinicians, and sometimes—even worse—to determine if something I might do as a psychiatrist may involve a medication that may seriously interact with another medication currently prescribed by another provider.

I now must "data mine" for my own notes, finding within those notes the most critical components needed for the reason I am seeing the patient now before me. I no longer have the same control I once had in designing my own paper notes, forced to follow the rules set by designers of the EMR.

EMRs—Masquerading as EHRs—Are Driving the Direction of Evaluation and Treatment in a Clinical Encounter and Fragmenting Care

While there is a concept for population health improvement, many IT systems were developed through a comprehensive acquisition process that included developing the strategic plan, business case analysis, and functional requirements with statements of work and deliverables. A request for proposals would be published and bids reviewed. The successful bidder would work with the contracting officer and designated technical reviewers. Among the IT products are the EMRs envisioned to replace all the EMRs in DOD, VA, and other federal facilities.

In the alcohol and drug program EHR prototype, Theresa, and Template, there was no question that the focus was on the patient and designed to improve the quality of the clinical encounter. Screen designs and flow captured current practice guidelines, and updates could reflect changes in those guidelines. Signs and symptoms that led to a diagnosis could be consistently captured and connected to specific treatment interventions. These were true EHRs of the 1980s–2000.

I've yet to work with an EMR where the EMR company and programmers have demonstrated interest in understanding the work I need to do and interaction I need to have to most effectively evaluate and treat a patient. I've become accustomed to being given a few training sessions to learn how the EMR works and then to use the EMR for documentation as it was designed to document. The process of patient care can be driven by the templates in the computer that have been given to me to use with various boxes to be checked or pull-down lists to be used along with free-text typing into specific fields, making sure the EHR format is followed while interacting with the patient implicitly becomes the guideline for the patient encounter.

This only reflects that those purchasing an EMR have not focused on improving the quality of a patient encounter and have not been concerned about patient's need for a continuum of health interventions over the patient's lifetime.

In the EMRs I've used since, it has been rare to find one designed to capture the clinical needs of both patient and provider—and to document as accurately and efficiently as was once possible on paper.

Today, under the military health insurance plan, I have some patients who are on military active duty and very aware that what I write in my EHR will never become integrated into the patient's military EHR. The military's EHR is not designed to allow me to view medical information that I need or would help me evaluate the patient or coordinate care with the military clinician who referred the patient to me for care. Although there are now standards for continuity of care documents that would be critical for an EHR, EMRs have not been designed for patient-centered continuity of care, or collaboration between providers in different places.

As both a patient and clinician, the EMRs I have seen used and have used have become distractions from time scheduled and needed for the direct encounter. The computer screen competes for attention, with the clinician shifting gaze between me and the computer. I do the same with patients. I now explain to the patient why I'm so engrossed with the computer—I need to find my last notes and find what I need to know today for decisions made before. I want to track the impact of treatment, treatment changes needed, or possibly information needed from other sources that may have impact on prior or new diagnosis and problems. Diagnostic and treatment changes then need to be discussed and documented. In this process, patients are sometimes surprised at how little information is in the system, and sometimes that data is incorrect. Patients are annoyed when they know they already gave information I am requesting to another clinician—but I am unable to find.

Contracting for What "Many Stakeholders" Say They Want—Not for What a Clinician Needs

What is the primary mission for an EHR? It's an opportunity missed if not thinking beyond the clinical encounter from the viewpoint of each provider. Each provider has a specific process where each patient is seen, collateral information is collected, medical record is written, and records are stored. Technology can be used to change from a large paper storage problem to a more efficient use of space with electronic storage. From a facility view: significant cost savings, but it's not helping improve patient care. Nothing has really changed from an inefficient paper documentation system—except I can't get the EMR as well organized as I could once with a paper record. I can't create my own "dashboard" so that both my patient and I can quickly visualize progress over time or developing problems.

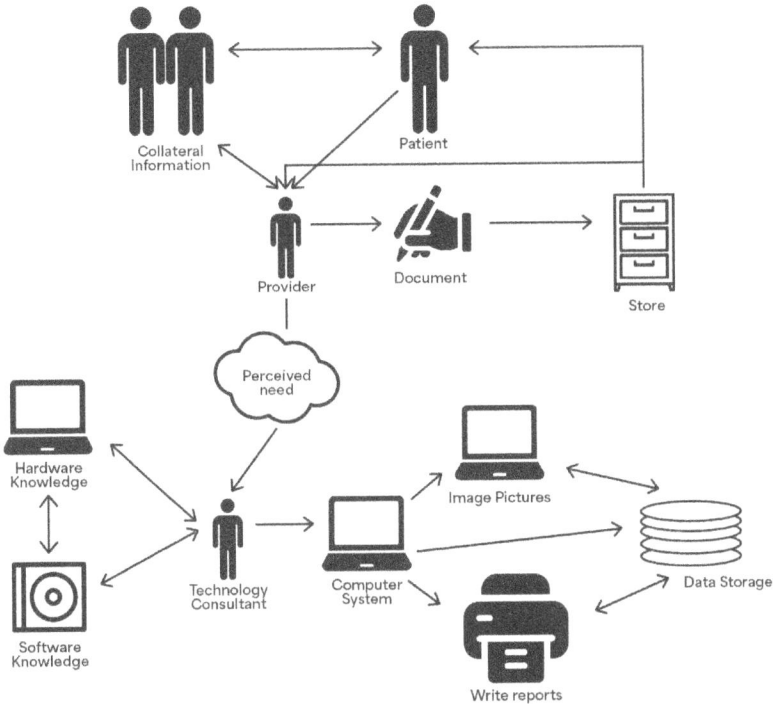

Figure 5.3

In decisions to purchase an EMR and converting from a paper-based medical record system to an electronic system, the paper process is reflected in the desired electronic system.

The resultant EMR is designed to the perceived need of the clinical experts who don't usually work together and don't question the "rules" behind the creation of the paper record. Once an EMR has been purchased for clinicians to use, they identify problems with the EMR, but the cost to change can be insurmountable. Electronically, the old inefficient paper system duplicates at every clinic, amplifying the chaos that was already in paper systems through EHRs adding documentation time necessary for each encounter.

This form-based EMR is just like the paper medical record except now it demands more time from the clinician to input required data necessary to generate reports needed by surveyors for various accreditation requirements or to meet quality care metrics.

Today, in one certified EMR, whenever seeing a patient, I must "order" a billing code at the same time I "order" medications—being audited on whether I've manually entered a proper date, time, and billing code, while I develop workarounds for my own notes so I can "data mine" the EMR for the clinical information I need to track each time I see the patient.

In medical record reviews of malpractice claims, dates, times, modifications, and a note accurately reflecting the encounter are all important. EMRs that are not designed for concurrent and accurate documentation at the time of the medical encounter can increase liability.

Entering the Age of Documentation for "Defensive Medicine"

With fear of malpractice claims, which can devastate the life of a conscientious clinician, came defensive documentation, increasing the time spent to document "everything" whether or not it's important for patient care. More words created clinical "data smog." There were so many written words not needed or necessary for the care of the patient, that it became difficult to find the critical data necessary to track treatment progress and changes in the patient's condition. More time was spent writing or dictating, adding to the clinician's workweek, with more documentation done after the patient encounter. Unfortunately, in a record review for a malpractice claim, date and time can be critical. Problems are created when notes are dated or written after the clinical encounter, and when clinicians realize something was forgotten or written in error, corrections or additions to notes related to the past can create more problems for the clinician in defending against a malpractice claim.

In malpractice claims, it is common practice to subpoena the medical record. Lawyers realized that different providers within the same and different disciplines may not have read each other's notes. Searching for inconsistencies and/or missed critical factors that lead to different conclusions by other providers, it could be possible to find the factors for a malpractice claim. It was commonly difficult to track through a large paper chart, which providers should have been working together to resolve the various patient diagnosis and problems that had been identified by different providers—and to identify gaps where there might have been a duty to provide care and a breach of that duty. A missed or incorrect diagnosis could be just the beginning.

When asked to review EMR records today, this problem has gotten exponentially worse. With paper documentation, brief and semi-structured documentation could be done within the time of the patient encounter and shared with the patient as standard practice.

When it came to potential malpractice claims, there were, and still are today, four essential factors: (1) a medical duty to provide care, (2) a breach of that duty, (3) the breach resulted in injury, and (4) damages resulted from the injury.

"Concurrent" and "collaborative" documentation was and continues to be a best practice. The patient–clinician relationship should be paramount in establishing the trust that the clinician is working in the patient's best interests and demonstrating care, concern, and competence. This can be the best defense against potential malpractice claims.

EMRs have not been designed for either concurrent or collaborative documentation—except when willing to have the patient observe how intrusive the EMR has become, interfering with a clinical encounter, and how difficult it has become for the clinician to find information important for the clinical encounter.

In the first chapter, the real intelligence of described EHRs and support tools was through software designed to connect accurate and necessary data with information immediately needed by a clinician to help the patient as much as possible within each brief encounter. With the uncertain accuracy of data in today's EMR, programming for Artificial Intelligence (AI) should take this into account.

> How this happened: Individuals and workgroups generating the functional requirements for a contracted EMR acquisition did not provide the specific clinical needs with sufficient clarity for a patient-centric EHR and a clinically more efficient and effective patient encounter.

Contract EMR Impact on Fragmenting Patient Care and Increasing Clinician Unhappiness

With one picture worth one thousand words, there becomes an evolution of many carefully developed contracts for IT hardware and software that can include provisions for an EMR, PHR, and patient portal.

This notionally captures systems and processes in place during the 1990s when DOD was developing the seven process steps for population health improvement. It illustrates a complex connection between various EMRs in different places, clinical decision-support tools, administrative systems, and individual process teams (IPTs) generating functional requirements to contract for either commercial or government "off-the-shelf" (COTS and GOTS) software.

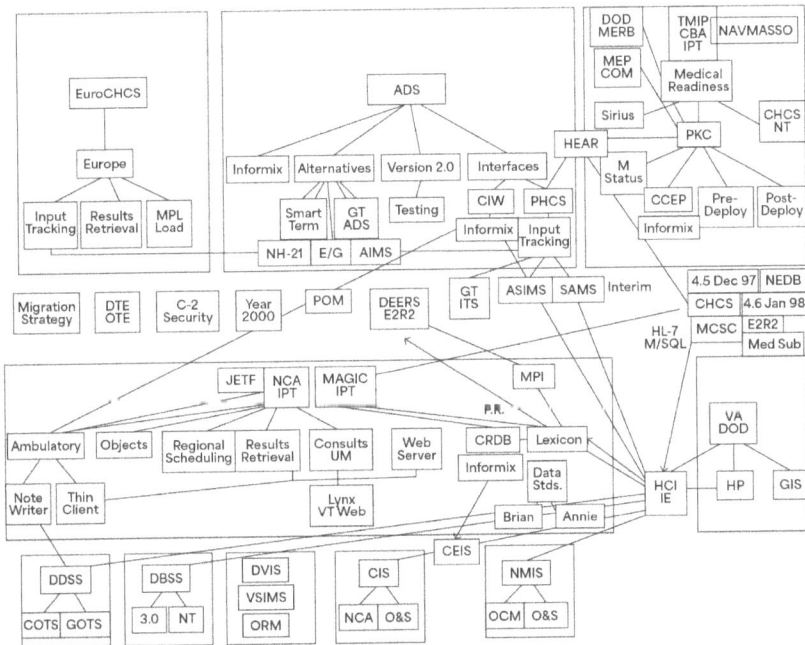

Figure 5.4

The Impact of Contracting for an EMR without Direct-Care Clinician Ownership: A Need to Document Patient Health Status Changes, Treatment Plans, and the Treatment Impact at Every Encounter

Without being able to think collectively about the array of interventions a patient might need worldwide, many discrete systems develop with different designs and different purposes. Once delivered, each proprietary system will be expected to perform per the contract under which it was created. If there is no functional requirement that systems interact with each other, they most likely will not.

I knew the meaning of most acronyms at that time. Behind each acronym were people spending time, with contract and financial support, for the projects identified in the diagram—to improve the healthcare necessary to improve and maintain the health of all DOD healthcare beneficiaries.

Where is the patient? Where is the clinician? What is the vision for healthcare? What is the pathway for many EMRs to become an EHR?

Underlying and Increasing Problems: Integrating Clinical Practice Guidelines into Clinical Care— Clinical Decision Support Tools Have Become Unusable for Use within a Clinical Encounter. Facility-Centric EMRs

Lost Has Been the Precision and Accuracy Necessary to Make a Clinical Diagnosis and Follow the Most Applicable Guidelines for Assessment and Treatment

For example, imagine a patient who has a potential opioid use disorder, while also being treated for chronic pain. They're only being seen for fifteen minutes:

If a patient has been diagnosed with an opioid use disorder (OUD), which of the eleven criteria listed in chapter 1 did the patient meet? With only two or three criteria needed for a diagnosis of a mild OUD, which ones were they?

What happens if a patient with a chronic pain condition, who is being given opioids for pain management, has developed tolerance to the opioid medication? They may experience reduced positive effect and withdrawal symptoms when not taking the opioid.

What happens if the two criteria for the diagnosis of OUD were the presence of opioid tolerance and withdrawal signs and symptoms? What if the patient is asking for more opioids, or has begun buying opioids on the street, perhaps now using heroin because it's cheaper and more accessible than trying to get a medical appointment?

Lost Has Been the Ability to Apply Highly Detailed, Diagnostically Specific Clinical Practice Guidelines within Clinical Encounter Time That Can Include an Emotionally Laden Interaction with the Patient

For example, Figure 5.5 is just the two-page pocket algorithm for the treatment of substance use disorder (SUD) from the VA/DOD website.

The DOD/VA's two-page algorithm for the treatment of SUD is so complicated that it's difficult to read at a glance. (Figure 5.5)

For this example: This is the two-page pocket card highlighting the three algorithms for the use of Chronic Opioids in the Management of Chronic Pain: Module A: Determination of Appropriateness for Opioids for Chronic Pain; Module B: Initiation of Treatment with Opioids; Module C: Maintaining, Tapering, Discontinuing, or Switching from Full Agonist Opioids.

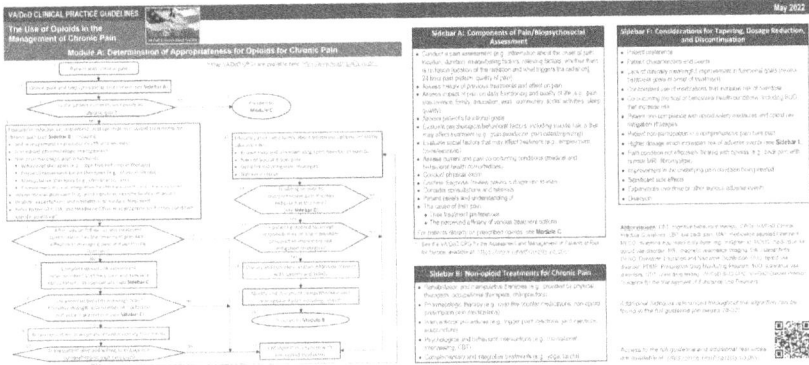

In a similar case, the DOD/VA's modules for the use of opioids in chronic pain management is so complex that it can't be easily read. (Figure 5.6)

For treatment, both methadone and buprenorphine have been approved by the Food and Drug Administration (FDA) for the treatment of chronic pain and OUD. It's packaged differently, but it's the same drug. Recommended therapies are different, but with commonalities not immediately obvious just looking at these algorithms.

What is the guideline if pain has been related to an infectious disease, like malaria, that has yet to be diagnosed or fully treated, but it's being treated now with opioids for pain and the patient meets the criteria for an OUD?

Today, there are over 2,700 clinical practice guidelines created by subject matter experts and various professional societies and organizations independently of a government. DOD/VA guidelines have been created through the support of the government convening recognized subject matter experts within the fields represented by the guidelines. With new knowledge coming from many peer-reviewed publications, these guidelines tend to become outdated within a few years, resulting in many guidelines being either archived or updated.

EMRs have not been designed to integrate this knowledge into the clinical needs for the patient care necessary within the short time of the encounter—nor designed to facilitate collaboration with the array of specialists who would be needed to realistically use these guidelines.

With high-quality data within each guideline, there are likely commonalities and efficiencies that might be an area for AI development. Using the logic present in the design of Theresa or Template, it would be possible, through screen designs, and help menus, to quickly implement best practices within a scheduled clinical encounter, with input from PHR data provided directly by the patient to implement and improve these guidelines over time. Clinical support tools and graphic displays would highlight the most critical data requiring immediate action.

EMRs have not been designed to improve healthcare results or correct the problems identified since 2000 in two sentinel joint Institute of Medicine and National Academy of Engineering reports: "Crossing the Quality Chasm" and "Improving Diagnosis in Health Care."

Building a Better Delivery System: A New Engineering/Healthcare Partnership

In the late 1990s, over seventy studies documented serious deficiencies and huge gaps between the care people should receive and care received. Physicians were being overwhelmed with an exponential increase in published knowledge, making it near impossible to assimilate new knowledge and guidelines into daily medical practice. Current healthcare delivery models focused on acute care for acute conditions, not the ongoing continuity of care necessary for patients with chronic conditions. Problems identified were lack of standardized performance measures that would help identify exemplary providers; that technology had not been used to help clinicians make better decisions; that physicians were operating in "silos" without access to the information and support needed to provide quality care. In the 2001 report of the Institute of Medicine and National Academy of Engineering, "Crossing the Quality Chasm," four key areas were recommended for change: (1) the use of information technology; (2) payment policies; (3) the development of best practices, decision-support tools, and an accountability system; and (4) professional education and training.

The Importance of a Thorough and Accurate Clinical Exam–and Making an Accurate Diagnosis

Since 1895, an international diagnostic vocabulary has been developed and updated every ten years, reflecting the increasing knowledge that has been used to improve diagnostic precision for every diagnosis; there are defined signs, symptoms, and other criteria.

According to the World Health Organization, "ICD [International Classification of Diseases] serves a broad range of uses globally and provides critical knowledge on the extent, causes, and consequences of human disease and death worldwide via data that is reported and coded with the ICD. Clinical terms coded with ICD are the main basis for health recording and statistics on disease in primary, secondary and tertiary care, as well as on cause of death certificates. These data and statistics support payment systems, service planning, administration of quality and safety, and health services research. Diagnostic guidance linked to categories of ICD also standardizes data collection and enables large-scale research."

Clinical data collection relies on a clinical exam that collects the signs and symptoms necessary to make one or more diagnosis or a differential diagnosis that defines the additional information necessary to make a more precise diagnosis. For any diagnosis, for accuracy, the specific signs and symptoms need to be documented.

An example: The following are eleven signs and symptoms, some number that must be present, in order to make a diagnosis of an alcohol use disorder (AUD).

11 Signs and Symptoms of AUD Since 1980

1. Alcohol is often taken in more significant amounts or consumed longer than intended.
2. A persistent desire or unsuccessful efforts exist to reduce or control alcohol abuse.
3. A significant amount of time is spent on activities necessary to obtain or use alcohol or recover from the effects of alcohol.
4. Craving or a strong desire of urge to consume alcohol; (replacing legal problems in 2013).
5. Regular alcohol use leads to an inability to meet responsibilities at work, school, or home.
6. Continued use of alcohol despite having persistent or recurrent social or interpersonal problems caused or exacerbated by the effects of alcohol.
7. Significant reduction of important social, occupational, or recreational activities due to alcohol use.
8. Recurrent alcohol use in situations in which it is physically hazardous.
9. Continued use of alcohol despite knowing a persistent or recurrent physical or psychological issue is likely to have been caused or exacerbated by alcohol.
10. Developing tolerance to alcohol, needing more alcohol to get the same effect, or that the usual number of drinks have less effect.
11. Experiencing withdrawal symptoms from alcohol such as shakiness, sweating, nausea, restlessness, racing heart, trouble sleeping, or seizures after stopping or reducing drinking.

If two to three signs/symptoms are present, AUD is mild, four to five present, moderate, six or more, severe.

The specific signs or symptoms present become the target for treatment. Not all with an AUD will have the same signs/symptoms, and treatment should be different.

"Improving Diagnosis in Health Care": A Moral, Professional, and Public Health Imperative

In 2015, the National Academies of Sciences, Engineering, and Medicine published the adverse consequences from inaccurate or delayed diagnosis. A diagnostic error was defined as "the failure to (a) establish an accurate and timely explanation of the patient's health problem(s) or (b) communicate that explanation to the patient." A root cause analysis behind diagnostic errors included "inadequate collaboration and communication among clinicians, patients, and their families; a healthcare work system that is not well designed to support the diagnostic process; limited feedback to clinicians about diagnostic performance; and a culture that discourages transparency and disclosure of diagnostic errors, which may impede attempts to learn from those events and improve diagnosis."

Those errors could seriously harm patients by preventing or delaying appropriate treatment, or providing unnecessary or harmful treatment, with psychological and financial consequences. Postmortem examinations over many years estimated diagnostic errors contributed to approximately 10 percent of patient deaths, perhaps the third leading cause of deaths in the United States, while hospital medical record reviews suggested that diagnostic errors accounted for between 6 and 17 percent of adverse events.

The National Academy stated that "improving the diagnostic process is not only possible, but also represents a moral, professional, and public health imperative."

With all exponential increase in diagnostically specific clinical practice guidelines and new information published every day, one constant has not changed: the time allocated for every patient encounter, with minimal elasticity to extend time with a specific patient having unexpected or new problems.

The EMRs I have used have not been designed to enhance my ability to make an accurate diagnosis, create a comprehensive treatment plan with measurable results important for the patient, improve the face-to-face time quality of the patient encounter, or make it easier to manage more than one specific patient problem per encounter.

I now create workarounds for the EMR in order to provide quality care; relying increasingly on stand-alone personal apps and internet searches to trusted websites while concurrently entering or transcribing relevant documentation into the EMR while interacting with the patient. Too much multitasking distracts from the attention and concentration that should be focused on the patient.

CHAPTER 6

By Gilbert Pant

A Technological Perspective

EHRs: How Did We Get It So Wrong?

You now know from a clinical perspective how we managed to get the goal of one provider to take care of one patient during one encounter in the most effective and efficient method so wrong. Here's how we got it so wrong from a technology solution perspective.

Let's start with first principles: The design of any system requires a full understanding and agreement on the requirements. Without it, nothing that follows will work. Historically, the leaders of industries defined those requirements: owners, managers, and administrative heads of the establishments. The established requirements for UK's government-owned National Health System (NHS), for example, were purely financial and administrative to maximize volume and output for a given budget. Private systems like hospitals and insurance companies often define their requirements in terms of financial administration of resources and billing. Within that perspective, the technology efforts of these establishments to produce desired results looked only at short-term solutions: hospital groups looking at the current year's bottom line, governments wanting to deliver before the next election. When attention turned to clinical support for medical staff, a system would frequently look at a single problem, at a single institution or provider, and often at a single episode.

Nothing highlights this better than the British government and the NHS during the 1980s under Tony Blair and the subsequent decades under different governments. This period achieved many things including, for a while, the massive reduction of waiting lists for hospital treatment but a considerable increase in spending. This led to greater thinking about efficiency improvements as a way of managing sprawling costs. At one point, the prime minister himself publicly called for patients to be able to book their own appointments, "providing a real choice, just like people can book an airline ticket for a flight." This resulted in the creation of a "choose-and-book system" that allowed a patient, once told by their general practitioner that they required specialist care, to search for a hospital appointment at a time and place of his choosing from the available choices. At the time, I required a specialist and was told by my generalist that he would like to refer me to a particular clinician who was an expert in the treatment of my condition. However, he was unable to make the referral directly. Instead, he could only pass me to the choose-and-book system. Having negotiated the new online system following an invitation to do this, I was still required to speak to somebody to select my appointment. When presented with a choice of six hospitals and various dates for the appointment, I asked, "Which would be best for what I have?" And the answer came back, "No idea, my love. We just make the appointments."

Fortunately, my doctor had suggested a specific specialist, and as he was one of those available, I was able to book my appointment. The system that was intended to speed things up not only cost significant funds to develop, but it also slowed the process down, because it looked at a single problem rather than the big picture and completely failed to take into account the way clinicians work and how they may specialize within their own fields.

This was all part of a process that had decided that improved automation would lead to better healthcare. But part of that process swept away a large volume of work being done by a number of smaller software houses in the United Kingdom that, working with clinicians, were developing clinical support systems alongside the hospital administrative systems. This happened because part of the reorganization to improve efficiency introduced major American patient management systems that, by being based on the US (insurance) model, were there to serve hospitals and institutions administratively rather than clinicians dealing with their patients.

Largely lost was the idea that by focusing on the patient–clinician encounter, a patient-based EHR system could significantly improve the quality of care at the same time as increasing clinical efficiency, by giving clinicians optimal support at the patient encounter.

This in turn would reduce stress and burnout of clinical staff, improve outcomes by reducing mistakes caused by a lack of information, and ensure that more time was spent with patients and less with keyboards.

The physical and mental health of an individual, a community, a society is related to the life cycle of people. It does not exist within the confines of a life cycle of a government or business plan.

Healthcare and the support of patient's health is a whole-of-life matter, as some chronic illnesses may be handled by various establishments over a patient's life, and even acute episodes may have an impact on health issues later. Whereas the administration of healthcare can be handled on an episode-by-episode basis, the relevance of clinical data to future healthcare cannot be predetermined, nor can the impact of care and conditions across disciplines.

This apparent conflict on the importance of data between patient administration systems and clinical support will not be resolved until that fact is realized and is embraced throughout the process of creating a fully functioning EHR.

Three primary reasons for the failure of EMRs from a technology standpoint are as follows:

- The belief that administrative needs and clinical data needs can be combined means that often clinical needs are not properly considered. Those who treat patients, the clinical staff, have had little to no input to the processes and design of the systems. They are rarely asked what they need, or if they are, it is at the level of their specialty or subspecialty and limited to the specific episode. Instead of seeking clinician input, they are simply told to use the systems they are given.

- When clinical-specific systems were originally developed, they were done so as a second level, in piecemeal fashion, by individual clinicians looking at particular problems in their area of activity with no regard for anything truly integrated and no regard for the lifelong health needs of a patient. And in the rare instances, when development is supported or carried out by clinicians, it's frequently met with opposition, no funding, and without any commercial oversight or interest.

- Forces the needs required for the physical and mental healthcare of humans into a one-size-fits-all approach. Batch processing of transactions in a medical establishment does not fit the "real-time" computing model that an EHR requires. The unique transactional nature of an individual patient–doctor interaction requires a doctor to have access to what is important to them at that precise time and in a form that fits with the nature of the clinical activity.

A glaring example of the disconnect between clinical needs and administrative needs was on full view at a meeting I attended at a large district hospital in the United Kingdom in the 1980s. A management consultant asked the gathered group of the thirty key managers to explain what their role was in the hospital. After hearing each one, he said, "Interesting, not one of you mentioned patients and treatment. Does this mean that nobody in this hospital thinks of treating the sick?"

Unfortunately, modern software development techniques encourage rapid development, prototyping, and the use of Agile techniques at the risk of not fully understanding or "discovering" the scope of the problem and its probable solution. This is a mistake. The problem must be thoroughly scoped, understood, analyzed, and created in an environment that allows integration of modules as they are developed. At present, the increasing development of systems addressing just one aspect of the problem, independent of all others, with no interaction with other systems across the clinical spectrum, simply results in more fragmentation rather than producing the integrated result which is needed.

Furthermore, that lack of standards and a framework within which to work creates risks of cyberattacks and breaches of security and data integrity that undermine the integrity of the whole. A project must be developed under the auspices of an agreed framework, or master architectural plan, with a thorough acceptance testing procedure on any software before it can be deployed.

The Challenging Complexities of the EHR

An EHR is not a single thing, because patients and their individual needs are not one thing, and clinical practice is not one thing. How many disciplines are there? Within each discipline, how many subspecialties are there focused on different aspects of the illness or treatment? How many new treatments and new diseases are there? How to best care for humans is constantly evolving as science evolves. What is the period of time that needs to be considered when searching for or making available the information that is relevant *today* for the clinician?

Then, there are the challenges of the distinctly different needs of a healthcare institution. How do the requirements of researchers funded by universities or pharma companies fit with the requirements of an overworked clinician in a busy underfunded public hospital treating the same types of patients? It seems impossible to capture it all. And how do you meet the needs of a highly specialized team dealing with extremely complex cases, as opposed to those dealing with routine cases in the same discipline?

In the early '80s, I had produced a system for the treatment of a type of cancer for a clinician who wanted the system to enable him to see one patient every fifteen minutes. "That's too much stuff," he said when I first showed him what I had developed. "I need it to show me what I need to see to minimize time waste and be able to quickly enter what I've done, so I can move to the next patient and be able to quickly access the information I need when I see that patient again.

I want to see the history and my plan in the first minutes, then leave as much time as possible for treatment."

The same day, in another hospital in the same city, an identical service told me they needed to collect much more data because research students need tons of information from the computer to be able to produce good research. "We need more data fields, not a bare bones system like that. What you're showing us tells us nothing."

Many systems have been developed to support clinical activity in certain areas. But have they been flexible enough to consider new research and new thinking? And how do these systems handle their data so that information transferred between clinicians becomes relevant to the recipient as opposed to being what the sender thinks is important? How are systems able to cope with chronic illness alongside short episodes of care?

And what of complicated cases where two different problems are handled at the same time between different clinicians separately? This was highlighted to me at one stage when I visited a busy university hospital to demonstrate both an obstetric system that managed the process of childbirth together with another system that managed the care and treatment of patients with early cervical cancer. The assembled audience was impressed with what we showed, until a professor said, "Interesting, but we deal with pregnant women who have cancer, so tell me which system would my colleagues use for that patient."

And how many combinations of illnesses and conditions, short and long term, might appear in the same person over a period of time?

Communication

Unlike other activities in the commercial world, modern health systems have failed to adequately communicate data between clinicians. Different data systems and different data requirements have made that job difficult.

The reason is simple. It's about misunderstood and underestimated complexity. Healthcare is about myriad processes, all using and reusing their own datasets, needing to communicate across disciplines and across institutions, and even across languages and countries sometimes in an ever-evolving information environment.

The nature of healthcare data has been misunderstood and underestimated at a time when everybody wants a quick fix. A clinician needs to see the whole of life, in order to save time and get things right *the first time*. The clinician needs to see information in a concise and relevant way, so important data is presented first, and detailed supporting information is available only if asked for. It's always about the person, not the clinical service.

Understanding the Data

Another mistake has been to misunderstand data and its value in information. A diagnosis is only useful if the supporting information that led to its definition is also available for review. How reliable is this information? How objective is it? How do you define subjective data such as "severe pain" so that its meaning is accurately transmitted to others? Is accuracy and reliability of a particular test, observation, or condition measurable and recorded?

And what is the effect on that over time? Does it keep its relevance, or does its relevance change over time or as a result of future research? Is its current relevance dependent on who is looking at it or a future event not yet recorded? Has it lost importance when it moves to another clinician in another discipline? Has new research and medical advances changed the meaning of data collected in the past? Has raw data been converted to coded data in an accurate way, or has accuracy been lost in the conversion? Has severity been noted and the source of the data checked?

How does the same data look when transferred between disciplines and clinicians? The patient record is the same, yet does a mental health patient who has cancer require the same health record as the cancer patient without mental illness? The patient is the same, and so is the patient's health record, but what of the needs of doctors and nurses caring for the patient? Has what needs to be seen been addressed properly, both in terms of comprehension and interpretation of what is vital? What are the views of the patient data that the clinician needs to see at this point in time? The clinician must inform the data to be inputted, managed, and extracted in a patient's health record. And that is simply not being done in today's EHRs.

Effect of Volume and Sheer Complexity

If one looks at all disciplines and subspecialties and the requirements of treatment plans and research data, the sheer volume of processes and data collected in each process almost defies comprehension. And until now, computing power was not adequate to be able to harness what is required and adapt quickly to changing needs and environments. Imagine the impact of HIV on sexual health processes when dermatology suddenly became a major hospital medical problem and spawned a whole host of other inter-related treatment regimens and protocols. And the need to treat COVID-19 patients? How are our clinical systems prepared for that? Systems need to adapt dynamically; you cannot ask patients to wait while systems are redesigned, tested, and redeployed. They need to be able to absorb change without any real time lost. Clinicians need to be able to request new ways to manage the data, to create new protocols, to ask their systems to help them interpret the data, and to collect new types of data without waiting.

The Needs Are Too Complex; It Just Can't Be Done

We believe it can be done. Before we go into the how it can be accomplished in today's world, let's review the evolution of EMRs that have offered insight from a technology perspective.

Template

Template, an EHR that I developed in the 1980s and 1990s in the United Kingdom and that was installed in thirty hospitals, was a PC-based departmental system that identified early on the need for the data collected to be clinician independent. A principle of the system was the idea that all clinical data is in fact the answer to a question. So, "answers" could be stored as coded or real data (to a greater or lesser extent) as the answer to a question. This then led to the idea that clinicians could be served by providing them with a view of the data in the form of answers to clinician's questions. This in turn led to the concept that clinicians could simply ask the computer system to provide a "view" of the medical record based on what that clinician wanted to see, with the ability to change the "view" dynamically by modifying their "template," and that the "template" could be tailored to meet each clinician's idea of how they wanted to work on a dynamic basis.

The idea of storing clinical data in this fashion opened the door to allowing multiple disciplines to share collected clinical data and thus open the door to a true health record.

Template made sales in the United States of America and in Australia, but eventually was abandoned after the company was sold, partly due to the lack of funding, lack of support from government (the single-customer NHS), and the opposition from hospital IT services to supporting a myriad small PC-based workshop-style systems proliferating through the hospital.

FACE in Mental Health Assessments

Clinical data is frequently highly subjective, and recorded handwritten paper in text becomes limited in usefulness when communicated to other clinicians. Much of medicine has subjective data and uses data that is far from precise in nature. Mental health is perhaps the absolute example of this where the recording of symptoms and assessments is totally subjective. A system called FACE (Functional Analysis of Care Environments)[4] developed in the 1980s was effective in capturing subjective mental health data in a form when a standard assessment could be transported between clinicians in an unambiguous way. Although still being used in the United Kingdom, it never won the support it deserved.[5]

A number of other systems were developed as result of duplication and collaboration; however, none of them succeeded over time as the proliferation of competitors and the politics of each individual hospital or group competing for dollars and patients meant that no single solution ever emerged, and certainly not one that served clinicians. Obsession with funding and financial control also meant that

---

4   Paul Clifford, "Assessing and Managing Risk in Mental Health Services: The FACE Risk Profile," 2017, https://www.semanticscholar.org/paper/Assessing-and-Managing-Risk-in-Mental-Health-%3A-The-Clifford/bd226bf99b86173c4d4828341865 0ff176b5fecb.

5   "FACE Toolsets Risk Assessment & Management," https://www.imosphere.com/care-and-support-tools/risk-assessment-toolset/.

systems focused more on throughput and numbers than quality of care, which it was thought was something that doctors would take care of regardless, as if their time was limitless. However, it was never understood that efficient care provision leads to better health, which would lead to lower costs overall in the long term.

Why Did These Systems Fail?

Lack of Funding and Support

Funds for the development of clinical record systems were limited. Mostly, they were clinician led, whereas corporate or government funding was available for the deployment of large-scale patient administration systems. These systems, mostly from the United States, supported an insurance model and provided for the deployment of resources and (in the United States) of billing. They provided an EMR, focused on billing; anything clinical got billed, so if it's billable, it tells you what was done, but it did nothing to help the clinician in the clinical care process, nor the patient in the management of their health in the long term. They were typically institution-based, so the data was not available if the patient needed care elsewhere. Furthermore, the sheer volume of ever-changing solutions required at the clinical level, and the divergence of ideas on what was needed, made consensus difficult and consequently the emergence of a commercially viable solution difficult to justify.

Large systems that had the support of senior management in clinical administration were viewed favorably. A small complicated system run by Dr. XYZ for their small group of patients was not, as it often created a disproportionate volume of work that would often go unnoticed and unrewarded. So, there were instances in which hospitals sabotaged the work of doctors trying to produce systems to suit themselves.

Lack of Scale and Long-Term Vision

If we talk of a health system that is patient based, we inevitably talk about whole-of-life systems.

Governments that fund healthcare projects are usually working based on a four- or five-year term before the next election. They come in, make changes, and expect those changes to bear fruit so they can claim success *before* the next election. A project requiring a longer view is not welcome as no politician will want to work hard to develop something that the next government takes the credit for. Add to that most officials responsible for health do not take the time to understand the problem from both the clinical and the technical perspective before declaring their intentions.

In private medicine, similar constraints apply where annual results are required, and the scale of the problem is too great for any one private, corporate provider to take on. It's clear that a solution to a true EHR system to cover whole of life, with the ability to manage care in any discipline in healthcare, both current and future, cannot happen under these circumstances.

Lack of Processing Power

Those that approached EMR in the shape of departmental systems, with each clinical specialty having its own system, quickly realized that massive amounts of processing power would be needed. In past decades, many PC-based systems, working in hospital departments, did a satisfactory job, but integration across departments needed the kind of computer power that was not readily available, particularly on limited budgets.

Unreliability of IT Systems

Computers were, and still are, unreliable. Even as reliability dramatically improved, a computer system that is used to provide critical care must be 100 percent reliable. PCs used to provide small services in departments on complex systems could not solve their lack of reliability.

Lack of Modern-Day Expert Systems

Early expert systems that have since evolved into AI systems were still in early development with little impact on a general EHR solution. What work is being done is limited (albeit to some highly useful areas like Scan interpretation) to very specific and limited fields.

No Internet or Cloud Storage

Internet was not available in the early development of EHRs and has not resulted in a serious re-think of how the internet, cloud storage, and cloud processing could be used. Some work is being done by some large tech giants (Microsoft and Google) but no universal solution has yet appeared.

A Notable Exception

In creating Theresa®, Henry Camp knew and planned for the fact that any component could fail and systems needed to adapt to the processing power present in any operating environment while using any hardware available. From the beginning, he worked with clinicians for the most efficient method of documentation, collecting only the data important for each clinical encounter, and to display the data necessary for the knowledge the clinician needed to have to assess and treat the patient. In the design, a "down system" equated to a dead patient, and for over twenty-five years, the system was up 99.999 percent of the time. Any patient record could be retrieved in less than one second with no degradation related to numbers of users on the system. Data was encrypted end to end with no security breaches over the twenty-five years. Adapting, interoperability, and using the latest technology were built into the initial design.

Why it failed appears more related to the fact that Medical Systems Development Corporation was a small company, without the social interaction and marketing of larger companies to the executive decision-makers responsible for purchasing decisions. Another paradoxical disadvantage may have been related to the relatively low cost of the EMR itself, when fully focused on supporting the desire of a clinician to take better care of a patient.

No Joined-Up Development, No Joined-Up Thinking

And finally, the vast variety of small solutions to each healthcare system has led to a proliferation of small-scale clinical solutions with little thought given to how these can be joined up. A successful EHR system requires joined-up data and processing.

To say that the connection between clinical and technical is essential to the development of an effective and efficient patient-centric EHR is an understatement. Both voices must contribute to the process of its conception and development. Through both our individual and collaborative work and experiences, Ken and I have combined our clinical and technical perspectives to imagine the ideal EHR. Here's how the meeting of the minds led us to this juncture.

Care has been taken to distinguish between an EMR and EHR, where the EMRs have been more facility-centric meeting needs of a facility providing patient care. Today, with the focus on a positive financial return on investment, the focus has become more on the reimbursable billing codes for which the clinician is responsible for documenting. EHRs have a different focus: patient-centric, where the person can become a patient at many places over a lifetime. The EHR documents the patient's initial health status and how that changes through any medical intervention over the lifetime of the patient. By design, some EMRs might have the potential to be an EHR, while others might not.

PART IV

The Cure for
the Chaos

CHAPTER 7

By Ken Hoffman

A Clinical Perspective

How we fix it: Reengineering healthcare for individual and population health improvement—providing the decision support a clinician needs in every clinical encounter; documentation specifically for the purpose of improving healthcare delivery and quality.

Unity of Mission, Vision, and Purpose for Medical Care for Individual and Population Health

With the many approaches to strategic planning and generating the functional requirements for an EMR, it's hard to think of one today designed to capture a process for reengineering and optimizing healthcare, while also prototyping an EHR designed purposefully for clinicians to like. The alcohol and drug prototype EHR used a methodology with an approach that seems unique today that also was at a unique moment in time with a program that had a unique connection to a military command that desired soldiers to return to duty quickly.

The alcohol and drug program had established clinical practice guidelines and clinician credentialing that were consistent throughout the worldwide network of outpatient programs located on every installation. All clinics were expected to achieve Joint Commission accreditation and used a consistent approach in both care and documentation to meet accreditation standards.

The focus for documentation shifted from the needs of "other stakeholders" to focusing on the clinician required to triage; comprehensively assess; create an integrated summary of problems and strengths; define a treatment plan that could be followed for outcomes and meeting treatment goals; and facilitate communication between clinicians whether as peers, in supervision, or in collaboration with outside specialists.

"Medical necessity" was defined by the care necessary that would most likely result in remission. This required a continuum of care and to eliminate fragmentation that normally occurs as patients are discharged from a higher level of care to a lower level of care.

We recognized that we needed to observe what clinicians actually did and what got in their way of taking better care of patients. This meant ensuring they were not burdened with documentation requirements set by others but had little or no meaning in the patient encounter.

We had to observe clinicians within their work environment taking care of patients and integrate an optimal practice model that required data support into screen designs and screen flows that lessened the time needed for clinical data input and displayed useful data throughout the clinical encounter.

We also had to obtain clinician consensus that we "had got it right"—clinicians like what had been designed and liked the resultant EHR.

Without this analysis, some EHRs had been created by programmers who intuitively understood the needs of a clinician for documentation that improved patient care and decreased documentation time. It was rare to see such EHRs.

It was equally rare to have program governance split between command and employer. The overall program had been a "command program" with medical having command over inpatient and outpatient treatment. Command had ongoing control for education and deterrence testing, leading to an integrated program that was community-based across a continuum of primary prevention, early intervention, and treatment services involving counselors. Within documentation, the importance of patient privacy and confidentiality was paramount. Meeting 42 CFR part 2 privacy standards while providing the sparse data commands required reflecting enrollment in the program, appointment times, missed appointments, and addressing command concerns related to performance, or a positive drug test in the command drug testing program.

Uniting to better population health is the path that can create the EHR that clinicians find indispensably helpful. And it could provide the data necessary to integrate the plethora of complex Clinical Practice Guidelines (CPGs) for meaningful use within a clinical encounter and improve outcomes as accurate data documenting evaluation, treatment, and outcomes are analyzed for the feedback necessary to improve access to care, and patient outcomes, with far less time required from the clinician for documentation. With direct patient involvement, documentation becomes collaborative and concurrent, with each encounter ending with a dated and signed note that was meaningful at the time it was written.

How we fix it: New and updated technology refocused to help take better care of patients while serving various business needs in the background–business grows because of better patient care–integration of applied research and shifting to precision medicine, as well as leveraging greater use of personal health apps that patients like.

At the national and senior healthcare executive levels, fixing this problem will require visionary healthcare leaders focused on health improvement in areas where it is possible to have pilot projects and initiatives. A starting point may be using IT to improve access to care and enhancing the quality of the patient encounter in underserved populations and remote areas currently outside the affordable costs or reach of current EMRs.

EMRs are not the same as EHRs but could be designed for the functionality defined for an EHR. PHRs seem to be already there. Health apps are available for any individual—whether or not they're a patient—to collect, display, and report health information of immediate importance to the individual and conceptually could be fed through a patient portal directly into the provider's EMR or EHR.

We might learn from nations developing a population health improvement system for healthcare—be careful not to "contaminate" those initiatives with systems that have been created for a facility-focused US healthcare system.

In 2013, the Boston Consulting Group presented a report at the World Economic Forum on healthcare and leapfrog opportunities. In that report, there was a graph mapping countries by cost per capita for healthcare and life expectancy with a clear division between countries on an ideal path for health improvement and those on the path to avoid, where much more money was being spent for little health improvement as measured by life expectancy. A few nations within the Organisation for Economic Co-operation and Development (OECD) were on the ideal path while the United States was on the far right of the Path to Avoid, spending far more per capita for healthcare while life expectancy was getting worse.

It would be among the nations within the ideal path or those with higher life expectancies that are looking for leapfrog opportunities possible with technology leveraged for population health improvement.

Today, I advise patients to keep a list of medical problems, medications, allergies, immunizations, and treatment history, especially for treatments that had been helpful in the past and medications that also had been helpful, harmful, or made no difference. Patients have been under the illusion that I have access to this type of information in the EHR I use, but that's almost never the case to the surprise of patients when I let them see what I have.

What I advise patients to have at least on paper has been the foundation for a CCD yet to flow through EMRs as patients see different providers in different settings using different EMRs.

Beyond my experience described in chapter 1, I have yet to see a similar business process building a methodology to better understand how money was being spent (activity-based costing) for work in healthcare that added value and work that did not. This approach helped define the functional requirements that would have led to an EHR supporting both individual and population health improvement. Whatever the reengineering methodology, there must be a unity of vision of the highest level of leadership and the desire for clinicians at the frontlines to have the technology that could improve accurate and efficient data collection for the decisions necessary for a diagnosis, treatment plan, and monitoring outcomes. There was potential to spend less time in documentation and more time in patient care, consulting with others, and in continuing education.

Whatever the methodology, it was key to have a unity of mission and effort at all levels of leadership and among all direct-care providers which might lead to lifelong patient medical records that are immediately helpful at any clinical point of service, assuring isolation and focus of the EHR to help take care of the patient, using a current recommended clinical guideline, and facilitating communication between providers. No EMR I've used today has been designed for this primary purpose.

Billing, scheduling, and other software necessary for managing the business can be developed independently of the EMR with that software able to extract only the relevant data needed for managing the business. This would return the EMR to its rightful purpose for accurately documenting care provided to a patient: focused on accurately and efficiently documenting work leading to a diagnosis, treatment plan, and impact of treatment in follow-up visits, integrating care being provided by others.

The greatest hope I have today, as a clinician, is the potential for personal health applications patients have found useful for their own health goals, such as those downloadable to smart phones and smartwatches. Allowing the connection for a PHR through a patient portal within an EHR seems to be an opportunity for quick collection minimally of vital signs and other parameters found in the health apps that would be most helpful to clinicians at the beginning of any visit. From this beginning, direct-care users must see the immediate impact of an EHR to support the clinician and patient within the time of a clinical encounter. The joy in medicine is seeing work producing the desired results that reflect the value or both time and work spent with the patient, with both gaining both knowledge and insight from an experience that's helpful to others.

Better systems would cut waste of resources, require less medication, reduce clinical errors, reduce resource waste, and would have enormous financial and political benefits, particularly in those countries where healthcare costs are becoming unsustainable. Looking at country-level data, the health expenditure to GDP ratio remained by far the highest in the United States of America at 16.6 percent in 2022, followed by Germany at 12.7 percent and France at 12.1 percent, according to the database OECD Health Statistics 2023. And yet in the United States, up to 40 percent of the populations has little or no healthcare coverage. And these figures show every indication of worsening as treatments become ever more expensive and clinical staff ever more overworked and insufficient.

Even a 1 percent improvement in clinical efficiency would be dramatic, but we believe that 5 percent or more would be possible over time, by increasing the time spent on patient care supported by a good clinical, patient-centric support system. In the United Kingdom alone, public funding for health services in England comes from the Department of Health and Social Care's budget. The department's spending in 2022 and 2023 was £181.7 billion. The vast majority of this spending (94.6 percent, or £171.8 billion/$220 billion) was on day-to-day items such as staff salaries and medicines.

So, even a modest 1 percent improvement in the United Kingdom could save £2 billion a year. That's equal to $2.3 billion.

And in the US, where healthcare costs exceed $4 trillion a year, the potential improvement in healthcare costs is staggering.

My Dream for What Might Be Possible in the Near Future with AI

I'm impressed with AI that listens to a clinical encounter, then synthesizes written documentation based upon what it has heard, matched to a diagnostic or treatment template to write a semi-structured note.

But, with such large variance today in EMR-level patient data, critical security breaches, and having "Framingham" quality medical records covering every patient's lifetime as a critical output, we must take critical first steps.

1. Enhance patient confidentiality and coordinated collaborative care

2. Assure the accuracy of patient-centered electronic documentation for AI data mining—only mining from accurate documentation for improving clinical guidelines

The ways in which AI can be utilized also varies. Each has a purpose and will be imperative to EMRs in the future.

Security AI: Available for use by every patient and provider.
Search through EMRs and programs connected to other data systems within and outside EMRs firewalls:

1. For the patient: who has access to their EMR record and the role of that person and who has used that access.

2. For the provider, related to a specific patient: what other providers are currently treating the patient or available for referral/consult within and outside the patient's insurance plan, with potential cost estimate understanding co-pays and deductibles.

Diagnostic AI: two-way analysis relating diagnosis with defined diagnostic criteria

1. For collected and documented signs, symptoms, duration, severity, inclusion, and exclusion criteria—to include collateral labs, procedures, and external data sources: display a list of potential (differential) diagnosis, with probabilities for each diagnosis in the differential, and data needed to more definitively establish one or more diagnosis.

2. For a definitive diagnosis, search through any available documentation to determine if the diagnosis meets diagnostic criteria, and if insufficient or incorrect, what additional data would be needed for a correct diagnosis—or if another diagnosis is the better/correct diagnosis.

Treatment AI: two-way analysis relating specific treatment interventions to outcomes

1. For finding commonalities and differences within the more than 2,700 peer reviewed clinical practice guidelines published today—with each medical specialty publishing guidelines, define core clinical functions common to all, common to some, and those uniquely different. Analogous to a tree: as the tree branches, it can relate a diagnosis to treatment commonalities and where a specific diagnosis has a specific treatment recommendation.

2. Independent of diagnosis, using baseline signs and symptoms, and changes in signs and symptoms to determine which treatment interventions have significant positive, negative, or no effect on signs and symptoms being followed.

"Personalized Medicine" AI: Using highly accurate and unbiased EHR patient records found within Diagnostic and Treatment AI, use only these records for "datamining" for patient-level personalized medicine matched to providers most likely to be most helpful at any given time.

1. Discovering where greater efficiencies might be gained in diagnostic evaluations for optimal treatment outcomes, with real-time changes to clinical guidelines based upon an ongoing analysis of this data.

2. Every patient has an individualized health plan while AI learns over time of potential improvements, possibly avoiding adverse events from patient experience. It is potentially able to better match patients with specific problems to providers with specific competencies.

3. With demographic, cultural, family, genetic, or other social determinants of health variables available, it can determine if specific interventions may have positive, negative, or no effects as a function of these "non-medical" determinants of health.

With this approach, not only will each patient have an accurate lifelong health record, but as a healthcare system, we will learn from our patients about the problems they are encountering and interventions that are helping—improving access to care at the right level of care, with the right providers, improving health outcomes, and lowering overall cost.

CHAPTER 8

By Gilbert Pant

A Technological Perspective

Creating the Right Environment

In the design of any new system, explicit understanding of the scope of the project is essential to its success. Today's EMR systems, however, have taken a fragmented view of healthcare by focusing on a single problem, or a single clinical environment, so that the scope becomes limited in functionality, time, and budget, as if reducing the volume makes for success.

In the scope design process, EMRs have also neglected to identify all stakeholders or understand their needs. Historically, two critical stakeholders, direct-care providers and patients, have not been included in the EMR design and development process. The result is EMRs that have been designed and developed without IT developers and IT specialists having a comprehensive understanding of the work a direct-care clinician needs to do in an effective and productive patient interaction nor a comprehensive understanding of the documentation a direct-care clinician needs to do effectively and efficiently in support of necessary decisions that must be made during the clinical encounter.

A successful EHR requires that clinicians can view any relevant information related to the client's recent, concurrent, or past encounters with other clinicians. That in turn requires clinicians to contribute relevant data to an EHR to ensure its availability to others—that contribution is motivated by the user actually receiving relevant data back. Clinicians working in a team or working individually will benefit from collecting data in a structured managed way that can be made meaningful to the treatment of the patient. Items like a record of drugs prescribed are managed by making the prescribing and the prescription the same action. So that ordering tests are linked to the collection of the results. Reduction in duplication of effort will likewise improve accuracy.

By defining the scope properly—the ability of a clinician to be able to treat their patient in the most effective way, with immediate access to the information that they need to know to properly treat their patient in the shortest possible time—with input from all stakeholders, the project can be broken down into definable and usable segments that, while useful independently, together provide the bricks with which to build the house.

An EHR that provides support to clinicians and aids in the healthcare of a patient must:

- Cover the whole of life. There is no start and end to an episode if the data collected may need to be seen in the future by another clinician.

- Remove all barriers between disciplines to ensure data generated by one discipline can be accessed in the future by another clinician in another discipline anywhere in the world.

- Continuously develop medicine (techniques, treatments, and diseases) alongside changes in the developing and aging human body.

In short, a person's health record must be capable of containing all available, necessary, and relevant health data for the whole of that person's life in whatever shape or form and in a way in which it becomes information to support the clinical process and make raw data available for processing by applications in the future for the personal care, in a multidisciplinary environment, by any clinician or paramedical service anywhere in the world.

To accomplish this, a clinician must, in the most efficient way possible, be able to record what they have done and why they have done it in such a way that data is collected in an unambiguous, efficient, and accurate way to the point where it can be used to provide information to clinicians in the future and understood in exactly the same way as it was intended.

This is complex. Language is complex. Text can be useful but ambiguous, especially if it requires translation. Even test results numerical in nature can be confusing when they cross borders. I recently had a blood test done in France but had it repeated in the UK as the clinician in the UK was "unsure" what some of the readings meant, yet this is pure science. Blood pressure in France of 12 over 8 becomes 120 over 80 in the United Kingdom, so data needs to be kept in a form that allows the information to be produced clearly. Opinion needs to be separate from fact, and the evidence supporting data should be kept. Types of data which are unambiguous (like blood group) should be kept distinct from subjective terms like "some pain." "Facts," however, can be a surprisingly nebulous concept with the potential to reflect individual bias, opinions given stated as fact based on individual reputation, or a common belief reflecting a collective interpretation of knowledge and beliefs within a culture or time. Use of understandable, universal terms should be qualified through a coding system that allows transfer of meaningful information. At present, the ICD has been an international effort to establish medical criteria that if present represent a diagnosis or description of a situation or condition that has an impact on health. But other coding systems exist and overlap, and the EHR can help by providing automatic equivalence between different coding systems.

While the US finally adopted the ICD 10 over two decades after it was originally published in 1990, the rest of the world is moving to the ICD 11. The ICD is the vocabulary of medicine, and a brief means of conveying the specific signs and symptoms upon which the diagnosis is based.

What we need is a patient record that contains all potentially useful data in an unambiguous form that for the patient accurately describes a subjective feeling—a symptom or objective state—a sign and for that form to be available to clinicians processed and presented in the most useful form *as required by that clinician at that time.* For this to exist, the system must allow the clinician to collect the necessary data generated by their interaction with the patient and to do so in the most efficient and comprehensive manner possible. After all, the doctor's role is to devote maximum attention to the patient's care, and the EHR should support that role rather than become a main source of distraction in the process.

This is a challenging task, but not impossible: some systems even thirty-five years ago came close to achieving this objective, although we have now lost our way in the rush to push out short-term quick profit solutions; but in today's world, massive advances in computing power and facilities make delivering a solution much easier.

Solving the Big Issues

The development of an EHR that meets the true needs of the health-care industry worldwide needs the backing of participants on an international and multidisciplinary level. Large corporations that will want to "do it all and reap the profit" will not aid in the development of the ideal EHR we envision. Initially, a collaborative effort is required, across countries and across disciplines, to develop a solution (or framework) that can be implemented over time, everywhere. This is possible, because the solutions would lend themselves to an approach where each clinical environment developed to help clinicians interlocks with those already produced to share data, rather than having disconnected solutions. Production of components on an open-source basis or agreed platform which conforms to an evolving standard could provide local solutions (as exists now) while contributing to a total integrated solution.

The development of components can be a profitable area for corporations to competitively work in. Another potential area for profitable enterprise is the provision of services related to open-source offerings once a framework has been agreed to and the mechanism exists to update and expand the framework in stages.

Buy-In

It will be up to any hospital group, government, or private hospital provider to join in the long-term effort. The development of components on a discipline-by-discipline basis will make it worthwhile to use even a small part of the system as the framework is accepted. The more users who work on the system, the more useful it will become.

The system lends itself to being adopted by a small-scale governmental organization with a limited population. That this organization draws support from government and academic institutions on an international basis would be key. Large international corporations would also have a role in providing services related to the produced products and in encouraging participation in the solution.

Fragmentation

Fragmentation is a known problem today; myriad EMR systems on different platforms with systems that do not talk to each other have contributed to the mess we now have. But if we can institute a framework on an international basis, which in turn becomes the "store" that holds all the standards relating to the final EHR project, then fragmentation would not be a problem as each component developed would adhere to established standards and fold it into the bigger picture.

Within that "store," intentional fragmentation can serve a purpose if it recognizes that each clinical activity, and in many cases, each clinician working in that environment has his own needs and methods of working, particularly in an academic environment where work may be leading edge or research related. This is only beneficial when viewed as part of the unfragmented whole and if all development confirms to an ever-growing list of standards and frameworks to allow the accumulation of stand-alone components to become the little bricks from which the whole solution emerges, like Lego.

Health Record Structures

The objective of any health record structure would be to create a lifelong health record that belongs to the patient rather than the institution, but which acts as a key or gateway to the data needed on a need-to-know basis to the treating clinician(s). Like today's health records, permissions would enable institutions to retain records for their own purposes; however, the main EHR would be held electronically in a "bank" and owned by the patient from cradle to grave in an electronic format, much in the same way that a bank holds personal bank records in electronic form, with access provided via a bank card. In countries such as France, this principle already exists with the *Carte Vitale* and the European Health Insurance Card (EHC) that holds core data and current medical notes and guarantees access to free healthcare. But the proposal is to make the card act as a gateway, or key, to all the data for the lifelong health record from which information can be presented to clinicians in the form that they need for each encounter, valid across all countries of the European Union.

The Clinical Object

All clinical information can be expressed and stored as the answer to a question. So, all answers to clinician's questions can be divided into classes where each class of question can store its answers in a form that makes sense for that class. Test results, symptoms, medications, treatments, and surgical procedures can all be classified as clinical objects with a defined structure to allow other systems to search for specific data and process it.

Clinical objects can then have qualifiers of various types, allowing the data to be interpreted as required according to date, accuracy, source, reliability, and so on. Data can be stored, but then be extracted and processed into information as is required by the clinician at the time they need it.

However, other data types may be accommodated such as dates (date of birth) and other objects such as radiology images, sounds, and 3D images.

Data Qualifiers

Qualifiers are essential to provide context to a specific data object. For example, qualifiers may be needed to define the nature of the data provided. A blood pressure reading provided by a patient needs to be repeated by a doctor during a clinical visit. Images and other data types may need qualifiers to better provide the information to indicate accuracy, thoroughness, and reliability.

Objective and Subjective Data/Text and Coded Data

Some data can be intended: the time taken to read and understand text, and lastly, the inability to process subjective data in volume. Some disciplines present this problem more than others; mental health diagnoses and symptoms are among the hardest to define accurately.

For such cases where assessments of the recording of symptoms like level of pain can only vaguely be assessed, recording and coding systems must be developed, adopted, and accepted across the discipline. Sometimes, coding systems overlap, but this in turn is not an issue, as equivalent coding systems can be provided.

Many successful (and even more unsuccessful) attempts have been developed in certain disciplines to enable the development of portable data even in cases of subjective data. The idea of portable data would mean that certain terms would have a clearly defined interpretation to remove ambiguity and therefore allow the meaning recorded by one clinician to be perfectly and unambiguously understood by any other person.

The principle is best explained by the analogy with the Beaufort score that allows a sailor to perfectly understand the sea state when communicated to by simply saying Force 7 gusting to 8 or 9. Just this definition accurately communicates both the wind strength and the probable wave height and sea condition.

In healthcare, the same principle can be applied to subjective conditions where no precise measurement can be made as in pain, discomfort, and discoloration levels. The treatment of mental health is fraught with subjective feedback from the patient and subjective interpretation by the clinician, but the portability and comprehension of this assessment data between clinicians is crucial. To assist in communicating this data based on a structure of common-ground identifiers, the FACE system was developed in the UK and implemented in several countries.

Coded data also has the advantage of being multilingual as the code can have a definition represented in any language.

Data Apps and Personal Use

Increasingly, individuals are using personal devices to collect clinical and health-related data that could play a role in an EHR, making way for apps to be developed.

Personal health data such as heart rate, blood pressure, weight, oxygen levels, and exercise levels could feed into the medical record. This data could then be fed back to the patient to provide updates to their health goals such as a consistent blood pressure level or weight loss or gain.

The Health Record and Where to Store It

It is a given that the highest level of data security would need to apply to medical records. Where data is stored at a personal level, it must be secure. Where data is used for research, it may be extracted, consolidated for that purpose but appropriately anonymized. This is not a new problem, and adequate rules exist for the management of this type of data.

PHI or IIHI must be maintained at the highest level of security, easy for the individual person to access personal data, but impossible for others who have no right to that data to be able to access that data.

Data Ownership

The health record must belong to the patient. This principle already exists in a number of countries, but all too often hospitals and doctors think they own the data they generate. If it is accepted that all clinical data belongs to the patient, notwithstanding the principle that a hospital or HCP may be given permission to keep a copy or part of it for their own purposes, then the solution becomes possible.

Where to Store It

If the patient owns their data, then the patient should be given a choice as to where to store it. Ideally, it should be kept by the patient as their lifetime health record in a format resembling a website, with a dedicated URL. The patient should be given a choice of who "hosts" their personal record, but all hosts would need to conform to "top secret compartmentalized" storage standards for PHI and be tested periodically by ethical hackers. Governments may wish to provide free storage in a secure environment in parallel with other authorized hosts.

Patients may wish to also retain a personal copy. This could be done via a micro memory chip and/or stored in a phone. This would be useful for remote locations where internet access might be difficult, or in the case of accidents occurring far from home where a personal copy may be the quickest way of identifying the person and their underlying health issues. Any personal copy on a chip or a phone could be synchronized with the master copy wherever internet access becomes available, like the synchronization of photos or emails on a cell phone now.

Data Editing

The adding or editing of the data stored would need to be done in a controlled manner, leaving a full audit trail of who changed it and why and how they changed it. A facility would be required for editing or enhancing data already recorded as well as the addition of new data. New clinical data could only be added by any person or source pre-authorized to do so.

Data collected and held efficiently would remove the need and possibility of duplication, and all data would need to be subject to qualification to ensure the transparency of its meaning, reliability, and subjectivity. For example, tests undertaken by patients at home may be less reliable than tests in a hospital environment, and the equipment used to undertake a test may affect the quality of a result. Similarly, assessments and results that include a level of opinion must be clearly identified as such.

All data must be stored with enough information associated with it for an evaluation to be done on its quality, not just at the time it is collected but also at any future time, when there is a re-evaluation of equipment or procedures, or where future tests and observation effectively negate data already collected. For example, if equipment is discovered to be faulty or if a lab technician has been discovered to have falsified results, it would be critical to allow the system to identify those data items to preserve the integrity of all the other data items in the system.

For the same reason, authors of data need to be separated from readers, and readers, or future data providers, must be free to edit or use current data to question, if not alter existing data. To a large extent, such rules developed to allow for data addition, comment or change, could be supported by algorithms and AI that is discipline dependent and transparent. Imperative to the process is the assurance that the highest level of data accuracy will guide how data is added and stored.

The ability to concentrate all data in a single health record that is shared by all will mitigate, and in some cases eliminate, the time spent by clinicians searching for information and repeating tests and processes as a result of inaccessible existing data. Once collected, the data remains available for any clinical need anywhere and at any time in the future without the need for wasteful repetition of past activity. A basic principle of automation is the ability of writing data once and reading it many times: a benefit that becomes lost or heavily diluted when data is spread over a wide number of discreet or incompatible systems within many different organizations and locations.

Data Management

Data management extends beyond the collection of valid, accurate, and pertinent data for which rules can be developed, maintained, and enforced by several methods and change control rules that already exist in many disciplines. Equally important is the security of the data once collected and the manner in which it is stored and protected from abuse. Protections developed and established in other disciplines could be applied to healthcare data. The owner of the data should remain the patient, insofar that anything that identifies the person would be stored on the patient's own record and only the patient holds the key to that data. How that is enforced may vary between countries due to local legislation and the unique variations in which healthcare services are delivered.

A country that depends on private health insurance would need different security for abuse of data by insurance companies than a country that offers government-funded healthcare, which is free at the point of delivery. While the incentive to hack health records for malicious reasons will exist, they will be less likely than that which exists for banks and credit card records.

The User's View of the EHR: The Clinical Dashboard

As previously stated, if an EHR's primary objective is to help the doctor take care of their patient, the clinician must have access to the most important and most relevant information from the EHR at the point of patient interaction to maximize the care they provide in the available time. This has two aspects with potential conflict if not designed carefully. The first is to ensure that the doctor sees and is alerted to what they need to know when in front of their patient. The second is to ensure that the information is available with minimal computer interface during the encounter, thus allowing the greatest amount of time spent providing direct care to the patient.

If the EHR is one thing, stored all together in a secure internet location—one place for each, one person—then the information to be extracted from the EHR by the clinician will vary according to the clinician's personal choice, their specialty, and the reason the patient is presenting. In practice, this means that the information that needs to be extracted for a follow-up appointment with a cancer patient will not be the same information that needs to be extracted for a mental health appointment to address a patient's anxiety, even when there are things that both need to see (such as current medication). Information shown will be drawn from the entire health record, but how it will be presented will be determined by the nature of that clinician's requirements.

There may be variation within a discipline such that a senior consultant treating complex patient problems in a teaching hospital with students present may want to see and collect information quite differently from a doctor working in an extremely busy general hospital with limited time per patient. Universally, the dashboard will need to capture within a few seconds the most important need for assessment and intervention within the brief time allotted for the encounter. The dashboard should make as much sense and have as much impact for the patient as it has for the provider. It must enhance informed consent and shared decision-making, assuring the patient understands the reason behind the diagnosis/treatment recommendations and agrees with the treatment plan.

Dashboard and Displays

The dashboard is merely the framework for what the clinician will see. The components of the dashboard are its "displays" that the clinicians may choose to include in their dashboard. The dashboard must be designed (for each use case) in such a way that the clinician can access what they need to see in the shortest possible time and is alerted to anything critical that might otherwise be missed. For this reason, the dashboard should be made up of movable displays that are relevant to the individual clinician, and which can be configured to meet the needs of the moment. Split-second decision-making may be paramount, and the displays must present information to allow that to happen, much as a pilot in a modern airliner needs to see what is relevant at that second, to enable him to make decisions quickly and correctly. Other types of information may also be highly relevant, such as "Is the patient getting better?"

Although the dashboard and its components will primarily be dependent on the discipline and the type of medicine that is being administered, the clinician will be able to modify what they see by arranging the components (or displays) according to what they need, either overall, or dynamically if the need of a particular patient requires it, or according to the data that is presented to them. An application in their own right, displays will fulfill one or more functions in varying levels of complexity, with multiple ages and layers. Many displays will only appear in certain specific clinical environments, and even then, the content of the display will be configurable to meet the needs of the individual clinician.

Display Topics

Medication

The medication display will show current medications and the diagnosis or symptom for which they are being taken. This display should include the option to add non-prescription medicine, show history, and alert the clinician to any contraindications or conflicts between symptoms and medication, both past and present. Any new symptom or diagnosis should force a check on current medication and similarly an alert.

Checklists

Certain types of medicine benefit from checklists and procedures. Prenatal care is one such type with its focus on routine checks to ensure that there are no potential issues and risks. For this purpose, checklists related to the timeline of the patient can form a key display to avoid errors. Checklists are also a useful way of recording that procedures have been followed and appropriate tests have been completed.

Episode History

Where the episode is following a course of treatment through to cure, a history of what has been done and the current point on the care path the patient currently is can be helpful if carefully presented. Where a plan is being followed, an episode history display can offer next steps and appointment or treatment schedules.

Medical History

A quick summary of past history is extremely important and can help to point to solutions of current symptoms: knee surgery following an accident would become important information for a patient suffering arthritic or knee pain some years later. AI can be used to help determine the importance of past issues without the clinician needing to read through many pages of previous history.

Vital Signs

Key vital signs and their patterns over time, such as weight and blood pressure, could provide certain information in graphic form quickly to physicians. Such information could include data from various sources. This could include data provided by the patient, which would require a reliability indicator on the information used to differentiate data provided by the patient and that collected in a strictly controlled clinical setting.

Alerts

Alerts are an essential display in many clinical settings and can draw information from anywhere in the medical record. An alert would be anything that either the system or a clinician has indicated as an alert. Such data may be pulled from past history or current tests, treatments, medications, and symptoms.

Care Plans

A care plan can offer a comprehensive plan for patients who are being treated in a multidisciplinary environment in which each clinician dealing with the patient must understand the overall care that the patient is receiving, often over an extended period. A care plan can also serve as recovery plan to provide seamless care from treatment to after care to long-term care as needed.

Appointment Booking

While appointments are key for the administrative management of a clinic, it contains nothing of clinical value beyond whether the patient received the time planned for evaluation and treatment. This should not interfere with or detract from the time a clinician spends with a patient to assess and treat the patient. Clinicians need interfaces to appointment systems so they can see the plans for their patients from patient administrative systems (PAS) specifically designed for that purpose. Linking various PAS independently would have the advantage of seeing appointments for the same patient in other hospitals or services and provide contact information for other clinicians involved in the patient's care. With this knowledge, permission could be obtained from the patient to view other medical records created by other providers past and present.

So, the dashboard for appointments should show more than the diary needed for appointment making.

Tests

Laboratory and other tests can be provided in both tabular and graphic form to highlight trends and anomalies that the clinician should see.

Images

Images and any comments associated with them can be displayed in the form most suited to the clinician.

Observations and Diagnoses

According to the specialty, this section may be more or less important and would need to be tailored to meet the requirements of the clinician and his specialty. But data shown here would need to provide links to the data that supported the information shown.

Social issues and family relationships (genetic issues). As above, this section might be needed for certain specialties, such as podiatry or mental health. It would need to be designed for the particular needs of the clinical specialty.

Display Needs Based on Area of Specialty

Acute Medicine

There is no one solution for a clinician managing acute illness requiring the clinical dashboard to manage the different practices between different specialties, the different level of specialization of a clinician, and the ability to pass information between clinicians, especially when patients are seen by other clinicians during a simple episode. Some acute medicine will be a single-visit episode, while others may require multiple visits and follow-up, but the main structure of the dashboard will follow standard guidelines, showing presenting symptoms and underlying health issues, medication, treatment and tests, and a history of the episode showing past appointments, and the plan for follow-up.

Not only will the dashboard be different for different specialties and subspecialties, but individual doctors within those specialties may also choose to see different things according to whether their focus is on volume or detailed care of complex cases. As the choices available to a doctor may change, the system must provide the ability to configure changes to processes promptly.

Acute medicine also allows for the use of rules, protocols, and AI to guide the clinician with checklists where appropriate and information on medication and treatments where conflicts may be identified and warnings given either between different medications or between medications suggested and other problems found in the personal health record (e.g., ibuprofen prescribed for a patient with renal deficiency).

Chronic Medicine and Care of the Elderly (including Geriatrics and Social Care)

Chronic medicine would appear in the EHR as an ongoing health condition to be alerted to any other clinician. In addition to providing a dashboard for the treatment of a chronic health condition, it would always appear as a summary of an underlying health condition when accessed by any other doctor. Comorbidities may require single-view displays of the patient to multiple clinicians at the same time.

Timeline history with medications may be especially important— with info on medication that are safer than others with potential contradictions highlighted—also monitoring mental acuity that may change as a function of health-promoting and medical interventions.

An opportunity arises to support "deprescribing" medication by ensuring regular reviews to remove medications no longer needed.

Surgical Applications

Each type of surgery may require a different format to the display and collection of data to be added to the clinical record. Surgical apps would be more likely to follow acute medical layout than chronic care, but there is potential for considerable data to be collected during a surgical episode, including anesthetics and recovery, which may be irrelevant except in summary to other clinicians. Changes in health from before and after surgery and potential complications that could develop immediately or later following surgery are critical information for the surgeon and for potential future encounters both related and unrelated to the surgery and the surgeon.

Infectious Diseases

The infectious disease dashboard would be similar to the dashboard needs of acute medicine with the addition of public health implications and reporting requirements. A known infectious disease will require different clinical information than a new, unknown disease. It would be the specialist clinicians who would decide what information the dashboard must provide.

Mental Health

Throughout this book, mental health has been one specialty where the EMR has had a detrimental effect on the documentation necessary to make a precise diagnosis and assessment. The inability for an EMR to pass accurate data between clinicians becomes potentially harmful. Psychiatric diagnosis generally includes specific signs and symptoms, level of severity, and other specific criteria. Related medical conditions may lead to a critical need for non-mental health specialists (e.g., drug overdoses and withdrawal, or heart and neurological problems related to medications or drugs, whether or not prescribed). Earlier EMRs had the capability to efficiently code standard assessments, and, like other specialties, allow for subjective and free-text comments that justified a course of action that is outside a recommended clinical guideline or reflect specific reasons for a non-specific diagnosis. Problems may arise related to a diagnosis made and treatment started based on inadequate data and bias, where the EMR itself may have created data connections the clinician does not know how to correct.

While mental health and SUD treatment programs have been subject to stricter rules about confidentiality, there have always been "break the glass" capabilities under emergency situations. The same higher privacy standards could apply for all patients with technology designed for that purpose. This type of information is a critical triage concern that involves the notification of others. What might be initially obscure needs clarity. This is where a patient-centric design becomes most helpful, perhaps at a lifesaving level for any medical specialty.

Emergency Medicine

This is a particularly difficult area to respond to as speed is of the essence. The implications are that the medical record may not be available either because of the location or the impossibility to identify an unconscious patient in the street or hospital.

In concept, an emergency room is to triage and immediately treat incoming patients with life-threatening conditions where intervention within the next few minutes may save a life. However, from the patient view, the emergency room may also be a resource of last resort, having a medical problem or concern but without quick access to any other care. Perhaps another person coming in is not so much a patient but someone seeking safe shelter in a warm place.

At the triage point, a person may have an altered mental status whose origins are not immediately known. Is there prescribed or unprescribed drug use? A stroke? The quiet patient in the corner may not appear to be in dire need of care but may be the one with minutes to live.

The information of greatest ER triage value would be data on an ambulance record, a patient who has a paper list of all medications taken, information coming from other people who come in with the patient, or the patient already wearing a medical alert bracelet clearly identifying a potentially life-threatening condition.

However, some support can be supplied if the patient has an emergency access card that can be used by paramedics, or some other ID that allows emergency data to be provided quickly via the patient's EHR, in an ambulance or upon arrival in an emergency department of a hospital. Each location will need to be addressed and support provided in the best way possible.

Ken saw the importance for the need of an EHR from his experience in New Orleans following Hurricane Katrina. Critical data may exist somewhere, but where? During Hurricane Katrina, over six thousand medical practices were underwater, with medical records destroyed. Patients came to the ERs, sometimes because they needed prescription refills, or perhaps seeking drugs—either could be at risk for serious/life-threatening withdrawal syndromes. However, all knew that medical data that once was, was no more, and critical databases, like pharmacy databases, existed but did not all interact should a patient have been filling prescriptions at different pharmacies on different databases. ER clinicians and staff had to more fully rely on the patient's subjective history and objective exam with labs possible in the ER for all necessary to reach a triage, treatment, and disposition decision.

EMRs as designed today have yet to make the transition to become an EHR. EMRs today can make the ER task more difficult, with variably helpful and potentially misleading medical data that might be stove piped within the EMR or not visible in other EMR records. Critically actionable data may not be found or if found, potentially misleading. For timeliness, is it better for an ER clinician to try and read through available existing EMR records, or start as if there were none?

Had the patient been hospitalized in the same place with the same EMR the ER uses, where is the most important information? The easiest found will be admission and discharge summaries and prior ER records.

Perhaps the patient has been in outpatient treatment with providers who are using the same EMR. Is this truly helpful in the ER? There may be hundreds or thousands of pages within the EMR, and somewhere within, related to the ER, highly important and actionable data. But where?

For quality of care or determining if care provided was for medically necessary reasons, Ken has had occasion to review EMR records. While he appreciates how much more legible an EMR is compared to variable legibility of paper records, he is always frustrated by how difficult it has become to extract medically important information. Recorded data can be conflicting: nurses' notes may identify different problems and concerns that might be in physician notes, and as different physicians and nurses provide care, data might also conflict or not be followed. Social work notes have similar problems. Actionable patient history may be outside physician notes but in social support services notes, therapist notes, or nursing notes. Blazingly clear is that the EMR has not facilitated coordination of care or communication between providers that might have resulted in a clear integrated summary of problems and interventions that will be tracked over time to determine the impact on patient health through a continuum of care—from patient home to inpatient or emergency services.

With EMRs designed to support documentation needed for billable reimbursable encounters for each provider involved with a patient, continuity and patient-centered care become second to care that results in reimbursement for the encounter that can include duplicating initial evaluations, labs, and procedures that were done elsewhere. Yet, this will still be unavailable as the patient becomes treated as a "new patient" for each different provider who now sees the patient for the first time—without access to prior history, problems, results of prior treatment, or treatment progress.

Rural and Remote

Many countries have an issue with communications and remote medicine. However, if the medical record has been accessed in advance, the dashboard can be designed to work offline on a pad or other device. The medical record can then be updated later when data access becomes available in a central location. Additionally, where remote clinics practice and data is offline, small populations are usually involved, allowing for the health records of the whole population of the location to be accessed and stored offline to be used by a visiting medic or medical team.

Important in this setting will be the relative non-availability of specialty providers and where more general providers or providers from related specialties may be called to provide specialty services for which minimal training has been provided. The ability to consult appropriate specialists and need for "on the job" knowledge and training may be critical for better outcomes. Rural and remote also translate to more care at or near the patient's home, sometimes requiring an urgent triage decision for medical evacuation or transportation over long distances.

EHRs supporting quality care in rural and remote settings provide a great improvement opportunity for EHR redesign that fully supports true patient-centered care, facilitating coordination and collaboration between general and specialty practitioners between all medical specialties and allied health services.

Community Health and Public Health

These services usually deal with routine clinics for PM or for other relatively minor acute episodes and would therefore require a dashboard that combines both the needs of acute medicine and rural and remote medicine. The compilation arises in that a single visit to a community service may generate more than one issue, and this needs to be taken care of by the system. But there would also be a need to report back to central public health services like vaccination campaigns.

Each public health department at city, state, and national levels may identify conditions of public health importance for which there is required reporting within various time frames—from within a day to monthly—and with varying need for individual identifiers, especially important if there needs to be clinical intervention that if not done becomes a public health threat (e.g., a case of measles, polio, or smallpox; potential new disease from travel elsewhere; child abuse or neglect).

An EHR designed to contain criteria necessary to identify suspected and confirmed diagnosis of reportable conditions, and, within its algorithms, also checks and triggers the provider to report conditions, enhancing the potential for rapid public health interventions. Healthcare costs can be lowered while life expectancy increased with public health, and EHRs can be redesigned to detect possibilities that there is a breakdown in public health: signs and symptoms related to contaminated water, sewage, or food, individuals with a highly infectious illness, or vaccine-preventable illnesses.

Well Services

Some healthcare involves patients who are not ill, such as obstetrics. In this case, the whole of the process, involving pregnancy and childbirth, simply follows a process where the objective is to identify and eliminate risk to the mother and child, a process different than most other health fields, and one that requires a dashboard of checklists and protocols to be followed. Eliminating errors by following checklists (like pilots in airplanes) helps to eliminate mistakes and errors and assures good outcomes. It also provides documentation to show that correct processes were in place and followed, which in turn reduces the risk of malpractice claims as documentation would be available to show that the clinician followed the guidelines correctly.

Obstetrics care highlights areas not well addressed in EHRs: health improvement and health promotion, which addresses nutrition, exercise, stress management, living environment, support in the neo-, peri-, and postnatal phases for mother, fetus, and newborn, with a lifelong medical record that should begin with the newborn to meet developmental milestones and quickly identify conditions for early interventions.

Multidisciplinary Health

One of the major problems facing clinicians is information regarding a patient with multiple simultaneous health issues. For example, a mental health patient with cancer or a cancer patient who becomes pregnant or a diabetic patient involved in a car crash. The universal EHR can create a dashboard for the attending clinician that provides a window into their patient's simultaneous health issues and direct contact with the other clinician(s) in real time where possible. Prescribing medicine would then be on a patient basis, not on an episode basis, eliminating errors through contraindications.

While today, the patient encounter tends to be one-dimensional—medications, physio/psychotherapy, social services—health and illness could and should be viewed in at least three dimensions: biological, psychological, and social/environmental, creating a more efficient treatment plan with better outcomes at lower levels of care.

The Patient's View of Their EHR

People would be able to access their own health records for a number of reasons. The most obvious would be to see and renew prescriptions automatically. In addition, there should be facilities for patients who are managing their own care to record data and communicate information to their primary care physician. In many countries (like France), patients have their own medical notes and are able to carry medical reports and radiology and lab results with them.

Modern devices that record blood pressure, heart rate, and the amount of exercise a person does would also be usefully interfaced with a personal health record. The interfacing of all the data supports the potential development of a lifelong tracking of health status and interventions: building a timeline history back to birth of medically important health status data and impact of various personal and medical interventions.

While the patient will have full access to the information in their health record, what the patient dashboard and its displays will look like is challenging to discern at this stage and will evolve as the clinicians' needs are fine-tuned and information is requested by the patient.

Emerging Technologies and the Future

In the 1970s and 1980s, health systems lacked many of the technical capabilities we have today. These current-day technical capabilities have created the existence of facilities that have dramatically reduced the cost and speed of processing data. AI is increasingly being used to assist in clinical care. In the same way that autopilot in aircraft assists pilots in guiding a plane to a safe landing, AI can assist clinicians in providing better and faster care by suggesting medication and treatments and highlighting elements of the medical record that should not be overlooked by a doctor.

AI is already healing through the analysis of radiology and other images helping to identify anomalies often in a more accurate and consistent way than traditional visual review by specialists. Similarly, lab results can be automated and managed by AI.

AI can also convert speech to text and (with caution) convert that text into useful processable data. AI is still in an early stage, and like all emerging and evolving technology, there are cautions to be taken when incorporating their use. A clinician may report, "No risk of diabetes seen," and AI has the potential to identify the text of "diabetes" as the patient's diagnosis.

Given what I understand about AI, and its potential use in healthcare, my greatest fear is that raw data and language in current EHRs might form a foundation where AI makes "discoveries" based on biased, incomplete, or misleading EHR records where data was not connected to criteria leading to a diagnosis or specifically guided decisions leading to treatment decisions. The design and workflow of the EHR have the potential to drive practice in directions that could create more harm to more people at a higher cost.

This could be prevented by following several steps:

- Helping design and develop the means through which high-quality data might be collected and identifying currently known data of low or incomplete quality.

- Collecting across various medically relevant data systems a composite health status timeline and interventions found in those systems that might be reviewed by both patient and current provider(s), allowing for a more precise and accurate update in a current medical record for immediate and future use.

- With the myriad of evidence-based clinical practice guidelines created by various specialty professional groups that are diagnostically specific, discovering the commonalities and creating overarching common guidelines that are possible to implement within a brief patient encounter, with decision support as needed and on demand.

- Discovering over time which interventions might work best with specific individuals—true implementation of precision-based medicine that goes beyond the limitations of "evidence-based" practices, unable to factor in the high number of variables that have direct impact on an individual's response to specific interventions and medications.

Big Data

The mass storage of anonymous medical data collected from large numbers of patients would allow organizations such as the National Institutes of Health (NIH) in the US and the WHO to analyze and produce research data on a massive scale as data would all be collected in the same way. We are far from the existence of universal access to healthcare big data. When we are empowered to move in that direction, it will be with an abundance of caution as to the existence of variable quality and uncertain biases that do not accurately reflect the information truly collected or the actions that were taken directly at the time of the clinical encounter.

Internet of Things

The internet of things would allow devices to automatically feed. People sometimes have their own blood pressure cuff, may go to a pharmacy or supermarket to measure their own blood pressure, or perhaps it is measured at all clinical encounters with any of many possible providers. Such data will be easy to collect, but the data would need to be held with a "reliability" indicator, so the clinician is aware of how the data was collected, such as a patient telling a clinician how much alcohol he consumes.

Cloud storage. The advent of cloud storage on a global basis now allows for a health record to be linked to a patient wherever he is or whatever he is being treated for. Increasingly, data is held in this form (electronic digital passports for example) so that a patient may have a complete health record at his disposal and made available instantly to any clinical or clinical service, granting a "key" to a medical institution.

Clearly, security is a key component, but selective or total access can be granted on the basis of "need to know" or "right to know" in addition to the granting of rights by a patient to an institution for care purposes.

The Way Ahead and How to Begin

To create the ideal, universal EHR, we must build a global standards organization that creates the framework in which to operate. This, by definition, must be a non-commercial group, although support from private enterprise, governmental organizations, and academia will have a role.

The new EHR standards (which we will call Health2045) will not replace other existing standards but will either adopt those existing standards or will provide mapping to and from those standards. The standards to be developed will include the following:

- Personal EHR record storage and access.

- Data object standards, for all classes of health objects that could be in the public domain for general use in the context of a continuity of care record/document.

This might be the core for a person/patient-controlled/owned health record that can connect with any of the EHRs in any facility with which the patient/person might interact. In this manner, health information flows with the individual while each facility and clinician can choose the "view" of the patient record it requires.

Dashboards or Clinical Views

Software developers will be invited to develop dashboards that use the newly created EHR. Dashboards should ideally be open source but not necessarily so, such that options and choices are available to users at varying cost levels. By creating dashboards for specific clinical situations, starting with certain common clinical require-ments, the EHR can be built incrementally over a number of years, while providing useful support for clinicians from an early stage of the product development.

If one year is required to establish the framework in an early stage to be useful, then applications could be available for use in many clinical environments within two years of the start of the process.

Companies and organizations would be able to use the experience and skills they have already acquired to adapt their software to fit the new framework, further reducing the time and cost needed to achieve useful results. Small countries may be easier places to develop solutions quicker, but parallel work would enable some economy of scale by the adoption of the agreed standards.

A useful dashboard component could be an individual's timeline history of health status changes as a function of personal health activities and medical interventions. Perhaps we could even leverage what dashboards exist today on smartphones and related health apps.

CONCLUSION

We need a new approach to EMRs that better understands the clinical support needed to optimize healthcare and utilize resources for the purpose of individual and population health improvement. With this understanding comes the potential for existing EMRs to transform and become EHRs.

There was once a bond with a healthcare system that allowed clinicians to focus on their patients, assuring the clinician had the most critical information needed for the encounter. Documentation was focused on the need to communicate with other providers and having the data needed for follow-up and decision-making.

The military healthcare system never questioned the need for an organized career-long paper medical record that went with service members and their families from post to post, from one medical facility to the next. The military alcohol and drug treatment programs exemplified a population health approach: providing the education needed for a healthy community; integrating into community health promotion programs; and providing best-practice treatment to achieve full remission on the first episode of care. Documentation supported continuity of care from inpatient to outpatient care world-side. The first EMRs described were building clinical practice guidelines into the screen designs and flows to improve the quality of the patient encounter. Independently developed clinical decision-support tools provided clinicians critical knowledge for immediate use within the clinical encounter.

In a critically missed opportunity, the importance of population and individual health improvement, a lifelong medical record, and documentation to improve communication—between clinicians for continuity, collaborative, and coordination of care—were not high priorities for the functional requirements for EMRs purchased, to replace the paper records.

The Korea case study highlighted the impact of what is possible when a healthcare system is created for individual and population health improvement, supported by accurate actionable data collected for the purpose of informed and shared clinical decision-making between patient and provider. It's a stark contrast to IT systems that had been purchased without understanding the clinical operational environment requiring efficient documentation for mission-critical use, nor the importance of having healthy and fit people able to deploy on a moment's notice.

At a critical junction and missed opportunity to use IT for health improvement, EMRs were designed and purchased to support a different business model. Healthcare became more entrenched in facility-based systems focused on return on investments and documentation more focused on justifying reimbursable medically necessary care. EMRs became designed for non-concurrent documentation, data fragmentation, collecting redundant, potentially conflicting, and inaccurate data with actionable data poorly displayed or not found. As designed, EMRs I now use interfere with time that could have been spent focused on the patient. Not designed to efficiently collect accurate data leads to problems related to misdiagnosis and inappropriate or delayed treatment.

If a fix is possible within the existing facility-based systems using various EMRs, a "back to basics" approach might reorient the systems to become patient-centered. Exemplary past EMRs were designed to be EHRs. Current EMRs have the same potential. Every individual should have a lifelong health record. Security policies with EHR implementation can allow data seamlessly shared with patient permission with any provider in any location involving the patient. With accurate, non-conflicting, and actionable data collected within each patient encounter, the business grows because of better patient care and new knowledge discovered while it provides care that is rapidly integrated into best-practice guidelines.

That can only be done through a whole-of-life health record that allows clinicians to have a total picture of the patient with a system that highlights what needs to be known at the time it is needed. This should be in parallel and share data with those systems that already exist, without collecting inaccurate data that could be misleading if included in a cradle-to-grave health record.

For the EHR we need, there must be unity of vision, mission, and purpose for a healthcare system.

Direct-care clinicians: Concurrent documentation with accurate and clinically important information had been the norm for excellent documentation prior to EMRs. Become an advocate for an EHR that would genuinely help provide better care and achieve better outcomes. Identify the positives and negatives of your current EMR. Is there anything that has helped improve the quality of the patient encounter? State the obvious specifics that must be part of functional requirements. For example, how fast and reliable should the EHR be? What is the most critical information you need prior and within the clinical encounter, for any decision needed by the end of the encounter? How much time do you want to spend documenting with data you may never see or use again?

Healthcare systems administrators: Think of the experience and results you desire for any patient coming under the care of your facility, and fully support the natural desire for clinicians to have the best interaction with the patient possible—the clinician clearly conveying concern, caring, and competence—supporting the natural desire for clinicians to learn what interventions are working best for which patients, and able to do more for each patient within every clinical encounter. Understand and support what your clinicians need from IT that will improve the results and outcomes within the time scheduled for each patient encounter, and have any EMR vendor demonstrate the ability to support your clinicians. Clinicians should be able to quickly perceive that the EMR helps rather than distracts from the quality of the patient encounter. Thinking altruistically, support the need or desire for patients to be seen by other providers at other facilities, and remember that your facility is one of many within the community. Assure that the EMR is truly an EHR with continuity of critical data able to be retained by the patient in a PHR and flow between any provider and facility involved in the patient's care.

Government and insurers: A patient-centered, problem-focused EHR, helping clinicians provide better care, also helps identify the care that is necessary to build, improve, and sustain individual and population health at the national level or within the pool of insurance policyholders. Healthcare costs must be paid through collected taxes or premiums coming from all covered individuals. EHRs with accurate data collected for the sole purpose of improving the patient encounter, in aggregate (eliminating PHI), will have the capability to shift health interventions to the best location, with the best provider team. Ongoing analysis of patient outcomes and knowledge gained from research and patient encounters will lead to interventions with better health-related results. Equally important, it becomes possible to identify interventions leading to harmful outcomes and costs that have no health benefit. It becomes possible to identify efficient and effective interventions, with better access to care, at lower cost. EHRs support continuity of care as an individual might need to move to higher or lower levels of care through various facilities and different providers with potential cost savings simply from having accurate data reflecting what is already known about the patient in other facilities.

Healthcare researchers and public health agencies: An EHR with clinically accurate and relevant data collected for patient care within any clinical encounter should allow near real-time background analysis detecting potential threats of public health importance, and with health researchers able to identify the most effective interventions while pushing improved or new guidelines into the EHR through updates. By triangulating other data sources with EHR data, syndromic surveillance can be done real time with interventions that may be best applied at an environmental, community, or individual level. We develop the capability to learn from the experience of the millions of patient encounters occurring every day through the consistently accurate data collected for each of those encounters.

We could be learning so much from all the patients seen every day, finding the pathways for treatment and interventions that result in best outcomes, which are tailored to the needs of each patient and linked to the provider who is most skilled to work with that patient. A patient-centered lifelong EHR designed to support the immediate needs of a patient seen in any facility by any provider becomes the gateway for personalized and precision medicine, enhancing communication between providers, avoiding preventable adverse events, and achieving optimal outcomes through practice guidelines quickly integrating knowledge gained from accurate data needed for the ten-minute clinical encounter.

GLOSSARY

ACLS (Advanced Cardiac Life Support)

AUD (Alcohol Use Disorder)

BAS (Battalion Aid Station)

BPS (Bio-Psychosocial)

CCD (Continuity of Care Document)

CDA (Clinical Document Architecture)

CDC (Centers for Disease Control and Prevention)

CHN (Community Health Nurses)

CME (Continuing Medical Education)

COTS/GOTS (Commercial/Government "Off-the-Shelf")

CPGs (Clinical Practice Guidelines)

CPOE (Computerized Provider Order Entry)

CPT (Current Procedural Terminology)

CTEAM (Center for Training and Education in Addiction Medicine)

DEA (Drug Enforcement Administration)

DNBI (Disease and Non-Battle Injury)

DOD (Department of Defense)

DRG (Diagnostic Related Group)

DSM (Diagnostics and Statistics Manual of Mental Disorders)

EHC (European Health Insurance Card)

EHR (Electronic Health Record)

EMR (Electronic Medical Records)

FACE (Functional Analysis of Care Environments)

FDA (Food and Drug Administration)

GEIS (Global Emerging Infections Surveillance)

GIFIC® (Graphical Interface for Intensive Care)

HCP (Healthcare Provider)

HCPCS (Healthcare Common Procedure Coding System)

HIPAA (Health Insurance Portability and Accountability Act)

HL7 (Health Level 7)

HRA (Health Risk Appraisal)

ICD (International Classifications of Diseases)

ICI (Imperial Chemical Industries)

IDEF (Integrated Definition Modeling)

IIHI (Individually Identifiable Health Information)

IMH (Instant Medical History)

IPT (Individual Process Teams)

KHF (Korean Hemorrhagic Fever)

MARI (Medical Aggregate Record Inquiry)

MHS (Medical/Military Health System)

MIS (Management Information System)

NHS (National Health System)

NLM (National Library of Medicine)

OECD (Organisation for Economic Co-operation and Development)

ONC (Office of the National Coordinator for Health Information Technology)

OUD (Opioid Use Disorder)

PAS (Patient Administrative Systems)

PHI (Protected Health Information)

PHRs (Personal Health Records)

SAMHSA (Substance Abuse and Mental Health Services Administration)

SDOH (Social Determinants of Health)

SOAP (Subjective, Objective, Assessment, and Plan)

SUD (Substance Use Disorder)

VA (Department of Veteran Affairs)

WHO (World Health Organization)

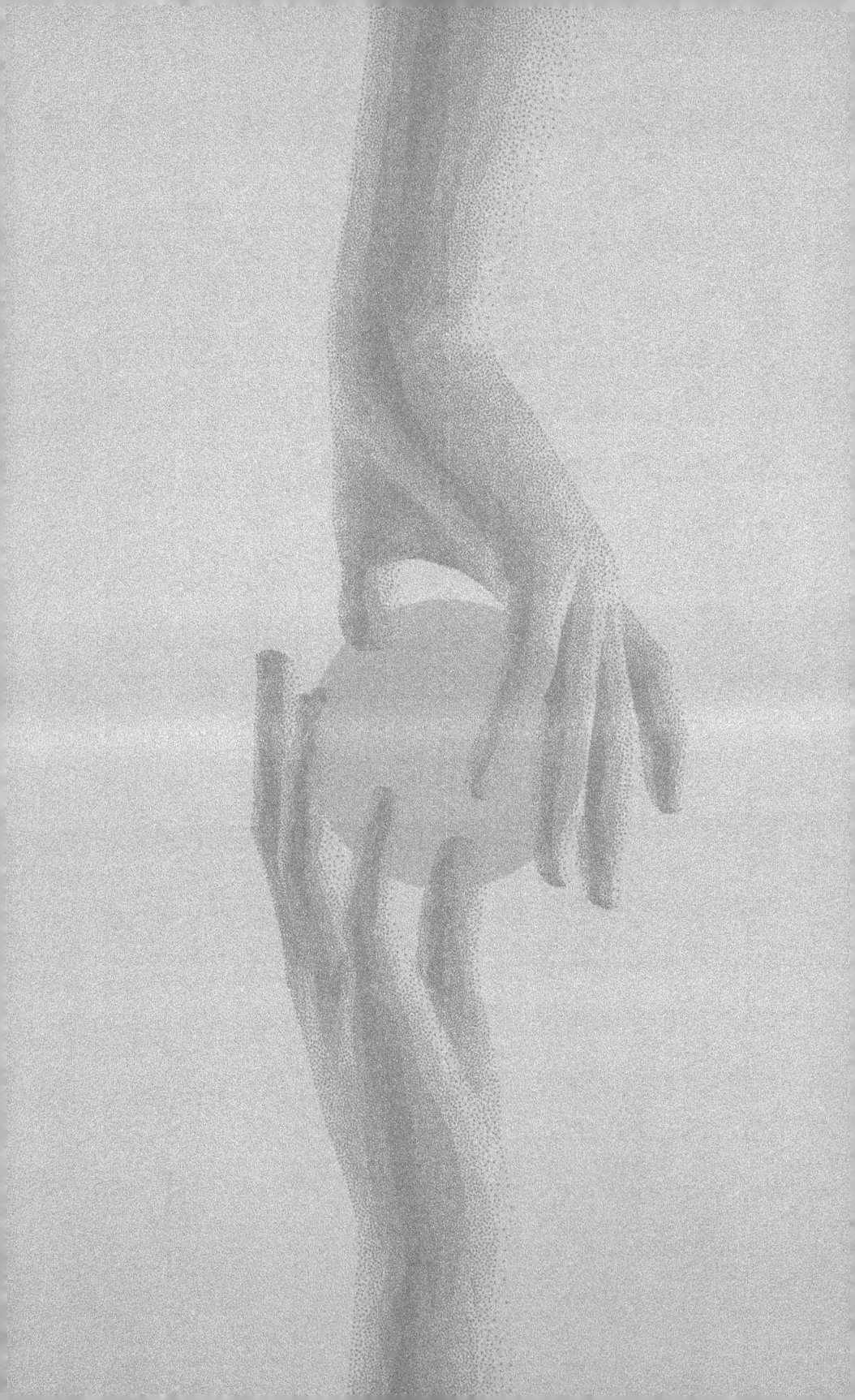

AUTHOR BIOS

Kenneth Hoffman, MD, MPH, Colonel (retired) Medical Corps, US Army

Ken grew up in New Jersey, graduated from Cornell University, and earned his MD from Eastern Virginia Medical School and MPH from Harvard School of Public Health. He completed his residencies in psychiatry and preventive medicine at the Maine Medical Center, Tripler Army Medical Center, and Madigan Army Medical Center. He is currently Board certified in general/addiction psychiatry and preventive medicine/public health.

Commissioned as an Army Medical Officer, he was trained as a flight surgeon focused on mental health. He developed an early interest in the prevention and treatment of addictive disorders, transcultural psychiatry, disease prevention, and health promotion. Ken has served as the 3rd Armored Division psychiatrist, director for the Center of Training and Education in Addiction Medicine at the Uniformed Services University, Preventive Medicine Officer for US Forces Korea, medical director for the Military and Veterans Health Coordinating Board, medical director for TRICARE Population Health Improvement, and the Army Surgeon General's alcohol and drug consultant.

Following his military retirement, Ken served as the senior medical officer at the SAMHSA Center for Substance Abuse Treatment and the Chief of Mental Health Services at the US Dept of State. Today, in private practice and consulting, he hopes the readers of this book might see the path he envisioned for the potential of a patient-centered EHR.

Ken lives in Maryland with his wife, Ann, with whom he has three children—Jay, Leah, and Phoebe.

Gilbert Pant MBCS, ACIB, CITP

Gilbert was educated at Emanuel School in South London and completed a banker's diploma in 1966, becoming a chartered banker with the Chartered Institute of Bankers.

Starting work on Midland Bank's (now HSBC) first computer system in 1963, he moved to Beirut with NCR in 1966, where he helped design some of the world's first banking systems in Beirut, Iran, Turkey, and elsewhere.

He founded his first company in 1971, a software house in Greece, with Alfa-Bank as a major customer. Later, in 1979, he founded one of the Middle East's largest computer companies, ITS, in partnership with the Kuwait Finance House in Kuwait (http://www.its.ws/).

Returning to the United Kingdom in 1982, Gilbert founded a healthcare software business that was active in the UK and the US, which was sold to Reuters in 1997. Since then, he has been involved in a variety of businesses and is a founder and major shareholder in a cloud infrastructure and web hosting company in the UK.

As well as being a chartered banker, Gilbert became an early member of the British Computer Society in the UK in 1965 and is a Chartered IT Professional (CITP).

He is a freeman of the Worshipful Company of Information Technologists in the city of London and a freeman of the city of London.

Gilbert lives in London with his wife, Nicole, with whom he has two adult children.